RealTime Physics
Active Learning Laboratories

Module 4

Light and Optics

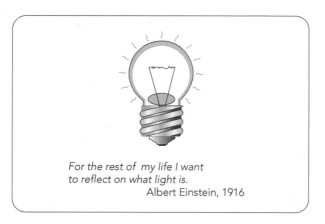

*For the rest of my life I want
to reflect on what light is.*
Albert Einstein, 1916

David R. Sokoloff
Department of Physics
University of Oregon

John Wiley & Sons, Inc.

EXECUTIVE EDITOR	Stuart Johnson
ASSOCIATE EDITOR	Alyson Rentrop
MARKETING MANAGER	Christine Kushner
SENIOR PRODUCTION EDITOR	Sujin Hong
PRODUCTION MANAGEMENT SERVICES	Aptara, Inc.
PHOTO RESEARCHER	Sheena Goldstein
COVER DESIGNER	Kristine Carney
COVER PHOTO	© CTR design LLC/iStockphoto

This book was set in Palatino by Aptara/Delhi and printed and bound by LSC Communications/Kendallville. The cover was printed by LSC Communications/Kendallville.

This book is printed on acid free paper. ∞

Founded in 1807, John Wiley & Sons, Inc. has been a valued source of knowledge and understanding for more than 200 years, helping people around the world meet their needs and fulfill their aspirations. Our company is built on a foundation of principles that include responsibility to the communities we serve and where we live and work. In 2008, we launched a Corporate Citizenship Initiative, a global effort to address the environmental, social, economic, and ethical challenges we face in our business. Among the issues we are addressing are carbon impact, paper specifications and procurement, ethical conduct within our business and among our vendors, and community and charitable support. For more information, please visit our website: www.wiley.com/go/citizenship.

Evaluation copies are provided to qualified academics and professionals for review purposes only, for use in their courses during the next academic year. These copies are licensed and may not be sold or transferred to a third party. Upon completion of the review period, please return the evaluation copy to Wiley. Return instructions and a free of charge return mailing label are available at www.wiley.com/go/returnlabel. If you have chosen to adopt this textbook for use in your course, please accept this book as your complimentary desk copy. Outside of the United States, please contact your local sales representative.

ISBN 978-0-470-76888-4

Printed in the United States of America

SKY10076563_060324

PREFACE

Development of the series of *RealTime Physics (RTP)* laboratory guides began in 1992 as part of an ongoing effort to create high-quality curricular materials, computer tools, and apparatus for introductory physics teaching.[1] The *RTP* series is part of a suite of *Activity-Based Physics* curricular materials that include the *Tools for Scientific Thinking* laboratory modules,[2] the *Workshop Physics* Activity Guide,[3, 4] and the *Interactive Lecture Demonstration* series.[5,6] The development of all of these curricular materials has been guided by the outcomes of physics education research. This research has led us to believe that students can learn vital physics concepts and investigative skills more effectively through guided activities that are enhanced by the use of powerful microcomputer-based laboratory (MBL) tools.

MBL tools—originally developed at Technical Education Research Centers (TERC) and at the Center for Science and Mathematics Teaching, Tufts University—have become increasingly popular for the real-time collection, display, and analysis of data in the introductory laboratory. MBL tools consist of electronic sensors, a microcomputer interface, and software for data collection and analysis. Sensors are available to measure many quantities including force, sound, magnetic field, current, voltage, temperature, pressure, rotary motion, acceleration, humidity, light intensity, pH, and dissolved oxygen.

MBL tools provide a powerful way for students to learn physics concepts. For example, students who walk in front of an ultrasonic motion sensor while the software displays position, velocity, or acceleration in real time more easily discover and understand motion concepts. They can see a cooling curve displayed instantly when a temperature sensor is plunged into ice water, or they can sing into a microphone and see a pressure vs. time plot of sound intensity.

MBL data can also be analyzed quantitatively. Students can obtain basic statistics for all or a selected subset of the collected data, and then either fit or model the data with an analytic function. They can also integrate, differentiate, or display a fast Fourier transform of data. Software features enable students to generate and display *calculated quantities* from collected data in real time. For example, since mechanical energy depends on mass, position, and velocity, the time variation of potential and kinetic energy of an object can be displayed graphically in real time. The user just needs to enter the mass of the object and the appropriate energy equations ahead of time.

The use of MBL tools for both conceptual and quantitative activities, when coupled with developments in physics education research, has led us to expand our view of how the introductory physics laboratory can be redesigned to help students learn physics more effectively.

COMMON ELEMENTS IN THE *REALTIME PHYSICS* SERIES

Each laboratory guide includes activities for use in a series of related laboratory sessions that span an entire quarter or semester. Lab activities and homework assignments are integrated so that they depend on learning that has occurred during the previous lab session and also prepare students for activities in the next session. The major goals of *RealTime Physics* are (1) to help students acquire an understanding of a set of related physics concepts; (2) to provide students with direct experience of the physical world by using MBL tools for real-time data collection, display, and analysis, (3) to enhance traditional laboratory skills; and (4) to reinforce topics covered in lectures and readings using a combination of conceptual activities and quantitative experiments.

To achieve these goals we have used the following design principles for each module based on educational research:

- The materials for the weekly laboratory sessions are sequenced to provide students with a coherent observational basis for understanding a single topic area in one semester or quarter of laboratory sessions.

- The laboratory activities invite students to construct their own models of physical phenomena based on observations and experiments.

- The activities are designed to help students modify common preconceptions about physical phenomena that make it difficult for them to understand essential physics principles.

- The activities are designed to work best when performed in collaborative groups of 2 to 4 students.

- MBL tools are used by students to collect and graph data in real time so they can test their predictions immediately.

- A learning cycle is incorporated into each set of related activities that consists of prediction, observation, comparison, analysis, and quantitative experimentation.

- Opportunities are provided for class discussion of student ideas and findings.

- Each laboratory comes with a prelab warm-up assignment, and with a postlab homework assignment that reinforces critical physics concepts and investigative skills.

The core activities for each laboratory session are designed to be completed in two hours. Extensions have been developed to provide more in-depth coverage when longer lab periods are available. The materials in each laboratory guide are comprehensive enough that students can use them effectively even in settings where instructors and teaching assistants have minimal experience with the curricular materials.

The curriculum has been designed for distribution in electronic format. This allows instructors to make local modifications and reprint those portions of the materials that are suitable for their programs. The *Activity-Based Physics* curricular materials can be combined in various ways to meet the needs of students and instructors in different learning environments. The *RealTime Physics* laboratory guides are designed as the basis for a complete introductory physics laboratory program at colleges and universities. But they can also be used as the central component of a high school physics course. In a setting where formal lectures are given we recommend that the *RTP* laboratories be used in conjunction with *Interactive Lecture Demonstrations*.[5,6]

THE LIGHT AND OPTICS LABORATORY GUIDE

The primary goal of this *RealTime Physics Light and Optics* guide is to provide students with a solid understanding of basic geometrical and wave optics. A number of physics education researchers have documented that most students begin their studies with conceptions about the nature of light, image formation, and interference and diffraction that can seriously inhibit their learning.[7–10]

RealTime Physics Light and Optics includes 6 labs:

Lab 1 (Introduction to Light):   The spreading of light from a point source, and the dependence of intensity on distance are first explored qualitatively and quantitatively. The differences between diverging and parallel rays (spherical and plane wave fronts) is explored qualitatively. Light reflected from extended and small objects is examined with a light probe. Finally, the use of light rays to represent the interactions of light with the surface of a transparent object is explored using a miniature light bulb (source of diverging rays) and a laser (source of parallel rays).

Lab 2 (Reflection and Refraction of Light):   The law of reflection and Snell's law of refraction are explored quantitatively using a laser to trace the paths of rays. Total internal reflection is explored using the laser, and the critical angle is measured. Dispersion and production of spectra and rainbows are explored qualitatively.

Lab 3 (Geometrical Optics—Lenses):   The mechanism of image formation by a lens is first explored qualitatively by examining the effect of a lens on the light from two point sources (miniature light bulbs). Misconceptions about the function of a lens are explored. Then the focal length of a converging lens is measured, and real and virtual image formation by this lens is examined quantitatively. Finally, focal length and image formation by a diverging lens are examined quantitatively.

Lab 4 (Geometrical Optics—Mirrors):   Image formation by concave, plane, and convex mirrors is explored qualitatively using two point sources, as was done with lenses in Lab 3. Focal length and image formation by a concave lens are examined quantitatively using a laser to trace the paths of rays. Then, image formation by convex and plane mirrors is explored in the same way. Finally, the images formed in the concave and convex sides of a spoon are analyzed.

Lab 5 (Polarized Light):   Polarized light is explored qualitatively using two pieces of Polaroid. A rotary motion probe with a Polaroid and light probe mounted on it is used as an analyzer to examine polarized light quantitatively, and develop Malus's law. The same apparatus is used to explore polarization by reflection. The direction of polarization of reflected light and Brewster's angle are then determined.

Lab 6 (Waves of Light):   After a qualitative examination of a single-slit diffraction pattern, the intensity distribution in the pattern is explored using a light probe mounted on a rotary motion probe. By sliding the rotary motion probe along a track, intensity and precise linear distances are recorded and graphed. Then the intensity distribution in a two-slit interference pattern is graphed. Modulation of the pattern by the slit width is observed, and a detailed analysis of the mathematical relationship for constructive interference is explored.

ON-LINE TEACHERS' GUIDE

The *Teachers' Guide* for *RealTime Physics Light and Optics* is available on-line at **http://www.wiley.com/college/sokoloff**. This *Guide* focuses on pedagogical (teaching and learning) aspects of using the curriculum, as well as computer-based and other equipment. The *Guide* is offered as an aid to busy physics educators and does not pretend to delineate the "right" way to use the *RealTime Physics Light and Optics* curriculum and certainly not the MBL tools. There are many right ways. The *Guide* does, however, explain the educational philosophy that influenced the design of the curriculum and tools and suggests effective teaching methods. Most of the suggestions have come from the college, university, and high school teachers who have participated in field testing of the curriculum.

The *On-line Teachers' Guide* has seven sections. Section I presents suggestions regarding computer hardware and software to aid in the implementation of this activity-based MBL curriculum. Sections II through VII present information about the six laboratories. Included in each of these is information about the specific equipment and materials needed, tips on how to optimize student learning, answers to questions in the labs, and complete answers to the homework.

EXPERIMENT CONFIGURATION FILES

Experiment configuration files are used to set up the appropriate software features to go with the activities in these labs. You will need the set of files that is designed for the software package you are using, or you will need to set up the files yourself. At this writing, experiment configuration files for *RealTime Physics Light and Optics* are available for the Vernier Software and Technology *Logger Pro* (for Windows and Macintosh) and PASCO *Data Studio* (for Windows and Macintosh). Appendix A of this module outlines the features of the experiment configuration files for *RealTime Physics Light and Optics*, as a guide to setting up configuration files for other software packages. For more information, consult the *On-line Teachers' Guide*.

CONCLUSIONS

RealTime Physics Light and Optics has been used in a variety of different educational settings. Many university, college, and high school faculty who have used this curriculum have reported improvements in student understanding of optics concepts. These comments are supported by our careful analysis of pre- and post-test data using the *Light and Optics Conceptual Evaluation* that have been reported.[10] Similar research on the effectiveness of *RealTime Physics Mechanics*,[11] *Heat and Thermodynamics*, and *Electricity and Magnetism* also shows dramatic conceptual learning gains in these topic areas. We feel that by combining the outcomes of physics educational research with microcomputer-based tools, the laboratory can be a place where students acquire both a mastery of difficult physics concepts and vital laboratory skills.

ACKNOWLEDGMENTS

RealTime Physics Module 4: Light and Optics could not have been developed without the hardware and software development work of Stephen Beardslee, Lars Travers, Ronald Budworth and David Vernier. We are indebted to numerous college, university, and high school physics teachers, and especially Curtis Hieggelke (Joliet Junior College), William Duxler (Pierce College), Gorazd Planinsic (University of Ljubljana), Julio Benegas (National University of San Luis), John Garrett (Sheldon

High School, retired), and Maxine Willis (Gettysburg High School, retired) for beta testing earlier versions of the laboratories with their students. We thank Mary Fehrs, Matthew Moelter, Gene Mosca, and Sharon Schmalz for beta testing the labs and for their invaluable suggestions and corrections in the final stages of editing.

At the University of Oregon, we especially thank Dean Livelybrooks and Stan Micklavzina for supervising the introductory physics laboratory, for providing invaluable feedback, and for writing some of the homework solutions for the *Teachers' Guide*. Frank Womack, Dan DePonte, Sasha Tavenner, and all of the introductory physics laboratory teaching assistants provided valuable assistance and input. We also thank the faculty at the University of Oregon (especially Stan Micklavzina), Tufts University, and Dickinson College for their input and for assisting with our conceptual learning assessments. Finally, we could not have even started this project if not for our students' active participation in these endeavors.

This work was supported in part by the National Science Foundation under grant number DUE-9455561, *Activity Based Physics: Curricula, Computer Tools, and Apparatus for Introductory Physics Courses*, grant number USE-9150589, *Student Oriented Science*, grant number DUE-9451287, *RealTime Physics II: Active University Laboratories Based on Workshop Physics and Tools for Scientific Thinking*, grant number USE-9153725, *The Workshop Physics Laboratory Featuring Tools for Scientific Thinking*, and grant number TPE-8751481, *Tools for Scientific Thinking: MBL for Teaching Science Teachers*, and by the Fund for Improvement of Post-secondary Education (FIPSE) of the U.S. Department of Education under grant number G008642149, *Tools for Scientific Thinking*, and number P116B90692, *Interactive Physics*.

REFERENCES

1. David R. Sokoloff, Ronald K. Thornton and Priscilla W. Laws, "*Real Time Physics: Active Learning Labs Transforming the Introductory Laboratory*," Eur. J. of Phys, **28** (2007), S83–S94.

2. Ronald K. Thornton and David R. Sokoloff, "Tools for Scientific Thinking—Heat and Temperature Curriculum and Teachers' Guide," (Portland, Vernier Software, 1993) and David R. Sokoloff and Ronald K. Thornton, "Tools for Scientific Thinking—Motion and Force Curriculum and Teachers' Guide," 2nd ed. (Portland, Vernier Software, 1992).

3. P. W. Laws, "Calculus-based Physics Without Lectures," *Physics Today* **44**: 12, 24–31 (December, 1991).

4. Priscilla W. Laws, *Workshop Physics Activity Guide: The Core Volume with Module 1: Mechanics*, 2nd Ed., (Hoboken, NJ, John Wiley, 2004).

5. David R. Sokoloff and Ronald K. Thornton, "Using Interactive Lecture Demonstrations to Create an Active Learning Environment," *The Physics Teacher* **27,** 340 (1997).

6. David R. Sokoloff and Ronald K. Thornton, *Interactive Lecture Demonstrations* (Hoboken, NJ, John Wiley ans Sons, 2004).

7. F. C. Goldberg and L. C. McDermott, "Student Difficulties in Understanding Image Formation by a Plane Mirror," *The Physics Teacher* **24,** 472–480 (1986).

8. F. M. Goldberg and L. C. McDermott, "An Investigation of Student Understanding of the Real Image Formed by a Converging Lens or Concave Mirror," *American Journal of Physics* **55,** 108–119 (1987).

9. B. S. Ambrose, P. S. Shaffer, R. N. Steinberg, and L. C. McDermott, "An Investigation of Student Understanding of Single-Slit Diffraction and Double-Slit Interference," *American Journal of Physics* **67**, 146–155 (1999).

10. *Active Learning in Optics and Photonics Training Manual*, David R. Sokoloff, ed., (Paris, UNESCO, 2006).

11. Ronald K. Thornton and David R. Sokoloff, "Assessing Student Learning of Newton's Laws: The *Force and Motion Conceptual Evaluation* and the Evaluation of Active Learning Laboratory and Lecture Curricula," *American Journal of Physics* **66**, 338–352 (1998).

This project was supported, in part, by the National Science Foundation. Opinions expressed are those of the authors and not necessarily those of the foundation.

Contents

Lab 1:   Introduction to Light /1

Lab 2:   Reflection and Refraction of Light /17

Lab 3:   Geometrical Optics—Lenses /33

Lab 4:   Geometrical Optics—Mirrors /53

Lab 5:   Polarized Light /75

Lab 6:   Waves of Light /91

Appendix A: RealTime Physics Light
and Optics Experiment
Configuration Files /107

Name_____   Date_____

Pre-Lab Preparation Sheet for Lab 1—Introduction to Light

(Due at the beginning of Lab 1)

Directions:
Read over Lab 1 and then answer the following questions about the procedures.

1. When is *ray* or *geometrical* optics generally applicable?

2. What is the meaning of light *intensity*?

3. What is your Prediction 1-1?

4. How will Prediction 1-1 be tested?

5. In Activity 2-2, what is the original source of the light measured with the light sensor?

LAB 1:
INTRODUCTION TO LIGHT

For the rest of my life I want to reflect on what light is.

—Albert Einstein, 1916

OBJECTIVES

- To discover how light spreads out from a point source

- To examine quantitatively how the intensity of light varies with distance from a point source

- To discover how light rays can be used to represent light as it spreads out from a point source

- To discover the origin of parallel rays of light

- To examine how we are able to see objects—how the light from objects reaches our eyes

- To compare the light from point and extended sources

- To observe qualitatively the interaction of light with the surface of a transparent object

OVERVIEW

Our ability to see the objects around us depends on light traveling from the objects to our eyes. Newton was a strong believer that light was made up of particles, and that it could be described by straight line rays drawn in the direction of motion. He could not conceive of light as waves. In the nineteenth century, a number of observations of light interference firmly established that light propagated through space as waves. Strangely, Einstein's theory of the photoelectric effect, early in the twentieth century, again established light as particles called photons.

We now know that light can be considered from both of these points of view—as particles or waves. When light interacts with objects much larger than its wavelength, it can be described by either waves or straight line *rays*. When it interacts with small objects—near the size of its wavelength—then a *wave* model is needed to accurately describe the interactions. Since optical elements like lenses, mirrors, and prisms are generally much larger than the wavelengths of light (which are of the order of half a millionth of a meter), the ray model—usually called *ray optics* or *geometrical optics*—is quite adequate. This laboratory will deal only with such situations.

INVESTIGATION 1: A THOUSAND POINTS OF LIGHT

When light is emitted or reflected by an object, each point on the object serves as a source of light. Let's explore how the light from a point source spreads out in space. The filament of a tiny light bulb, while not exactly a point source of light, is small enough to be a good approximation.

Prediction 1-1: Light from a point source spreads out in every direction. Based on the definition of *light intensity* as the *light power per unit area perpendicular to the direction in which the light is propagating,* how do you think the light intensity from a point source will change as you increase your distance from the point source? Be as quantitative as you can.

To test your prediction, you will need

- miniature light bulb
- power supply
- holder for light bulb
- computer-based laboratory system
- *RealTime Physics Light and Optics* experiment configuration files
- light sensor
- meter stick

Activity 1-1: Light Intensity Around a Point Source

Let's look at light coming from a point source at different positions around the point source. To do this, you can use a light sensor that measures the light *intensity** incident on the surface of its light detecting element (called a photodiode). Light *intensity* is associated with the *brightness* of the light.

1. Mount the light bulb in its holder, and connect it to the power supply.

2. Connect the light sensor to the computer interface, and open the experiment file called **Point Source (L01A1-1a)** to display intensity.

3. Use the meter stick to position the light sensor 2.5 cm from the light bulb filament, pointed straight toward it. Move the sensor slightly side to side until you get a maximum reading. Then, record the intensity reading in Table 1-1,

---

*Note that your light sensor probably measures *illumination* in a unit called *lux* rather than intensity in watts/m^2. The unit *lux* refers to the amount of light emitted by a specified source of white light. For our purposes, illumination in lux measures the same quantity as intensity, where 1 lux = 1.46 mW/m^2.

below. Now repeat this for four other positions around the bulb, all 2.5 cm away from the filament. Record these values in the same column.

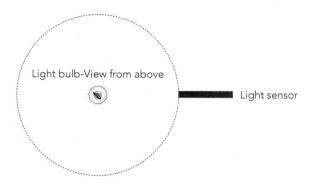

4. Repeat for 5.0, 10.0, 20.0, and 40.0 cm away from the filament. For each measurement, remember to move the light sensor slightly until you get a maximum reading.

Table 1-1

Intensity measurement	Distance from light bulb filament, R				
	2.5 cm	5.0 cm	10.0 cm	20.0 cm	40.0 cm
1					
2					
3					
4					
5					
Average intensity					

5. Open the experiment file called **Intensity vs. Dist. (L01A1-1b).** A data table and graph window will appear on the screen.

6. As you enter the data for Average Intensity vs. Distance (R) from the bulb in the data table, a graph will be plotted. **Adjust the Intensity and/or Distance axes,** if necessary, to display the graph most clearly.

Question 1-1: Describe qualitatively how the intensity measured by the light sensor appears to vary as the distance from the light bulb filament increases. Does the intensity increase, decrease, or remain the same?

Question 1-2: Try to explain your observations in words based on the way the light spreads out after it leaves the light bulb filament. (Remember that light intensity is light power per unit area perpendicular to the direction in which the light is traveling. Over the area of what surface does the light from a point source spread?)

7. Find the mathematical relationship between Intensity and Distance (R) from the light bulb using the **fit routine** in the software.

8. **Print** the graph with the fit included, and affix it in the space below.

Question 1-3: What mathematical relationship did you find between Intensity and Distance (R) from the light bulb filament? Try to explain this based on the way light spreads out from the filament of the bulb. (*Hint:* The surface area of a sphere is $A = 4\pi R^2$, where R is the radius of the sphere.)

Now let's examine how light can be represented by *rays*. In addition to the materials used in Activity 1-1, you will need the following:

- laser
- blackboard eraser and chalk dust
- transparent ruler

Activity 1-2: Diverging and Parallel Rays of Light

In geometrical optics, light propagating in space is represented by straight lines called *rays*, drawn in the direction in which the light waves are propagating. Light *wave fronts* are always perpendicular to rays.

In the previous activity you examined the intensity of light from a point source. Now let's look at how rays can be used to represent the light from a point source propagating through space.

1. In the diagram below, sketch 5 light rays beginning on the filament of the bulb and through the points 1–5.

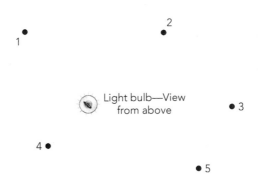

REALTIME PHYSICS: LIGHT AND OPTICS

Question 1-4: Describe these rays in words. How are they drawn as they move away from the filament?

Warning: The laser beam is intense enough to burn the retina of your eye. Never look directly into a laser or let the laser beam shine directly or reflect into anyone's eye.

Now observe the light from a laser shining across the room. It may be helpful to suspend some chalk dust in the path of the laser by pounding a blackboard eraser close to the path of the laser beam.

Question 1-5:   Can you see the laser beam very well without the chalk dust in the air? Why does the chalk dust make the beam more visible? Are you actually viewing the light coming from the laser directly, or is something else happening?

Question 1-6:   List all the ways in which the light from the laser appears different from the light from the small light bulb.

Question 1-7:   In particular, compare how the light from the laser spreads out as compared to the light from the light bulb.

2.   Based on your observations, sketch below some rays of light leaving the laser.

Question 1-8:   Would you expect to find any rays through points 1, 2, or 3? In what ways are the rays you drew different from those you drew from the point source in (1)?

Comment: The rays from a point source are radii spreading out from the source. These are called *diverging* rays, and the light is said to be composed of *diverging spherical* waves. (The wave fronts are spherical surfaces.) The rays from a laser are essentially parallel—they hardly spread out at all. They are

---

called *parallel* rays, and the light is said to be composed of *plane* waves. (The wave fronts are planes.) Are there any other situations in which the light from a source can be represented by parallel—or *nearly parallel*—rays?

Prediction 1-2:   Imagine a circle 5 cm in diameter on the ground. It is a clear day, and the sun is high in the sky. Are the rays reaching the circle from the sun more like the rays from the small light bulb (diverging) or like the rays from the laser (parallel)?

To test your prediction you will need

- miniature light bulb far away from your lab table
- power supply
- holder for light bulb
- meter stick
- transparent ruler
- small paper disk

Activity 1-3: Rays from a Distant Source of Light

1.   Measure the distance from the light bulb to your table:_____m.

Question 1-9:   How does this distance compare to the diameter of the paper disk?

2.   Hold the paper disk up with its surface facing the light bulb.
3.   In the diagram below, use the ruler to carefully draw several rays from the filament of the bulb to the disk.

Light bulb

|

Disk

Question 1-10:   Was your Prediction 1-2 correct? When light from a very distant source of light is incident on a relatively small object, is it best represented by approximately parallel rays (and plane waves) or by diverging rays and spherical waves?

INVESTIGATION 2: LIGHT FROM THERE TO HERE

Exploring the Light Around You

Let's look at the light coming from various objects around you. To do this, you can again use a light sensor.

First a couple of predictions:

Prediction 2-1: Look at the surfaces of objects that are near you. Choose two surfaces, and decide which one looks brighter to you. Describe the surfaces and your prediction below.

Prediction 2-2: Place a white and a black sheet of paper next to each other on the table. Which looks brighter to you? Which do you think reflects more light?

To test your predictions, you will need the following:

* computer-based laboratory system
* *RealTime Physics Light and Optics* experiment configuration files
* light sensor
* sheets of black and white paper

Activity 2-1: Light Sources and Reflected Light

1. Connect the light sensor to the computer interface, and open the experiment file called **L01A2-1 (Reflected Light).** Light intensity readings should now be displayed on the screen.

2. Test your predictions by pointing the light sensor at the surfaces of the objects and sheets of paper. Be sure that the two sheets of paper are right next to each other. In each case, hold the sensor so that its end is the same distance above the surface—about 5 cm. Be sure that the sensor is pointing straight at the surface and that you are not casting a shadow on the surface with your hand or the sensor.

Question 2-1: Was the object that appeared brighter also the one for which you measured the larger intensity? Describe your observations.

Question 2-2: Which piece of paper appeared brighter? Is this also the one for which you measured the larger intensity? Describe your observations.

3. Use the light sensor to determine where the light from the surface of the white paper is coming from.

Question 2-3:   Where did the light originate? What is your evidence?

Question 2-4:   Do objects like your body or the table top *emit* visible light? How can you see objects that are not *sources* of light? Where does the light come from?

Question 2-5:   Explain the differences in light intensity observed with the different objects in Questions 2-1 and 2-2 based on your answer to Question 2-4.

Light Spreading out in Space

You have seen in the previous activities that an object is visible either because it is a source of light or because it reflects light incident on it from a source. The brightness of an object seen by reflected light depends on the intensity of the light incident on it, and how effectively it reflects this light.

To better understand how we see objects, we need to know how light interacts with an optical component like a lens. And to better understand this, we must see what happens to light after it leaves a source or is reflected from the surface of an object.

We want to see what happens to the intensity of light as you move away from an object. First we'll look at an extended object—a sheet of paper—then at a small circle of white paper on a black background.

Prediction 2-3:   If you use the light sensor to measure the change in intensity of reflected light as you move away from the sheet or the small circle of white paper, compare the intensity changes. Will they be different for these two objects?

To test your predictions you will need

- computer-based laboratory system
- *RealTime Physics Light and Optics* experiment configuration files
- light sensor
- sheets of black and white paper
- small white disk
- large light bulb and holder

Activity 2-2: Light from Extended and Small Objects

1. Place the sheet of white paper on the table. Set up the light bulb so that it is about 1 meter above the paper, slightly off to one side and aimed at the center of the paper.

2. Set up the software to plot intensity vs. time by opening the experiment file called **L01A2-2 (Light from Objects).** This will display the axes that follow.

Note: If the intensity ever reaches a maximum value, and doesn't increase anymore when the sensor is moved, then the sensor is *saturating*. The intensity is too large for it to read. Change the sensitivity of the sensor or move the light bulb higher or away from the center of the paper until the sensor no longer saturates.

3. Practice plotting a graph as you slowly and uniformly move the sensor upward, away from the paper, starting 1 cm above the surface of the paper with the sensor pointing straight downward. Move the sensor at a slow uniform rate so that it reaches a height of no greater than 20 cm in the 20 seconds of graphing. You should start moving the sensor at the moment the graphing starts, and reach the desired height as close as possible to when the graphing stops. Repeat until you get it right.

4. When you have a good graph produced with uniform movement of the sensor, move the data so that the graph remains **persistently displayed on the screen.**

Question 2-6: Where does the light reaching the sensor originate? Describe the complete path it takes from the source to the sensor.

5. Now replace the white sheet of paper with the small white disk in the center of the black sheet of paper. Again graph the intensity starting at 1 cm above the disk and moving up to 20 cm above, at the same speed as before in (3). Repeat until you get it right.

6. Sketch your graphs on the axes that follow, or **print** your graphs and affix them over the axes. Fill in the intensity scale. Also, fill in an approximate corresponding distance scale below the time scale based on your estimate of the number of centimeters moved per second.

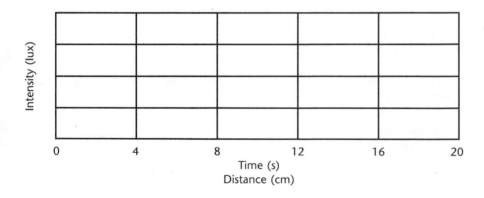

Question 2-7: How does the graph for the small white disk differ from the one for the whole sheet of paper? Did your results agree with your prediction?

Question 2-8: For which object does the intensity fall off faster from its initial value? Was it much faster? Explain this result.

Question 2-9: Would the rays close to the white sheet of paper be more like the rays you saw in Investigation 1 for the point source or for the laser, i.e., diverging or parallel rays?

Question 2-10: Would the rays from the small white disk (when viewed from far away) be more like the rays you saw in Investigation 1 for the point source or for the laser, i.e., diverging or parallel rays?

INVESTIGATION 3: UNDERSTANDING LIGHT WITH RAYS

Much of geometrical optics deals with optical elements like lenses, mirrors, and prisms. You will examine the interaction of light with these in later labs. Many optical elements are constructed of transparent materials. Here we want to observe *qualitatively* what happens when parallel rays (plane waves) and light diverging from a point source (diverging rays) are incident on a flat (plane) surface between two transparent materials, for example, air and glass. In the next lab, you will examine these interactions quantitatively and arrive at the laws of *reflection* and *refraction*.

To carry out these activities you will need

- laser
- glass rod
- miniature light bulb
- power supply
- holder for light bulb
- semicircular transparent chamber filled with a *slightly* cloudy liquid
- blackboard eraser and chalk dust

Activity 3-1: Interaction of Parallel Rays with the Surface of a Transparent Object

1. On a separate sheet of paper, set up the laser and glass rod so that the light is incident on the flat surface of the transparent chamber at an angle around 45°, as shown below. (The only function of the glass rod is to spread out the laser beam vertically so that it is more easily seen on the paper.)

2. Observe the light after it strikes the surface, both the light reflected from the surface (the chalk dust will be helpful for this observation) and the light transmitted through into the liquid.

3. Outline the shape of the chamber on the paper, and also sketch both rays.

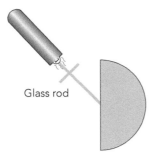

Glass rod

Question 3-1: Describe what happens to light rays when they are incident on the surface between two transparent materials.

Prediction 3-1: What will happen to light from a point source when it is incident on the surface of the transparent chamber? Sketch your prediction for three rays from the point source incident on the flat surface at the three different points indicated on the diagram below. Show both transmitted and reflected rays.

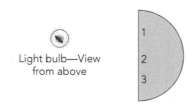

Light bulb—View
from above

Test your predictions.

Activity 3-2: Interaction of Diverging Rays with a Transparent Surface

1. Tape the miniature bulb to a separate sheet of paper so that it is pointing up from the paper. Place the transparent chamber in front of the bulb, as shown above.

2. Observe the reflected and transmitted light.

3. Sketch several rays on the sheet of paper.

Question 3-2: Did your observations agree with your predictions? If they didn't agree, explain why.

Question 3-3: How could you use the beam from a laser to predict what will happen to just one of the diverging rays from a point source?

HOMEWORK FOR LAB 1: INTRODUCTION TO LIGHT

1. What are *rays* and how are they used to represent light?

2. Give two examples of light sources best represented by *diverging* rays.

3. Give two examples of light sources best represented by *parallel* rays.

4. The intensity at a distance 5.0 cm from a point source of light is 25.0 watts/m². What is the intensity at a distance of 2.5 cm from the source? 15.0 cm from the source?

5. Compare the change in light intensity with distance from a point source to the change for a source of plane waves (parallel rays). In which case is the change faster? Explain.

6. When you look at a photograph, where does the light that reaches your eyes originate?

7. On a clear, sunny day, you are holding a magnifying glass close to the ground. Sketch several rays from the sun incident on the magnifying glass. (You need not continue the rays through the glass.)

Magnifying glass

8. You have a small *frosted* light bulb. There is a nonreflecting surface behind the bulb.

   a. The bulb is turned on. Describe how the light intensity varies as you move further and further away from the light bulb.

   b. Now the bulb is turned off, but light from a lamp is shined onto it. Describe similarities and differences from (a) in the way the light intensity varies as you move further and further away from the bulb. Explain your answer.

9. You hold a small lighted bulb in front of a window. In the diagram on the right, sketch several rays from the bulb that are reflected at the surface of the glass.

Light bulb

Name_____   Date_____

Pre-Lab Preparation Sheet for
Lab 2—Reflection and Refraction of Light

(Due at the beginning of Lab 2)

Directions:
Read over Lab 2 and then answer the following questions about the procedures.

1. Define *angle of incidence*.

2. How will the direction of the incident ray be determined in Activity 1-1?

3. How will the transparent chamber be used to examine refraction of light?

4. What is *total internal reflection*?

5. In Investigation 3, why will both white light and red laser light be shined through a prism?

LAB 2:
REFLECTION AND REFRACTION OF LIGHT

Somewhere over the rainbow way up high,
there's a land that I heard of once in a lullaby.
—E. Y. Harburg and H. Arlen, 1939

OBJECTIVES

- To examine the reflection of light at a surface between two transparent materials quantitatively

- To understand the law of reflection

- To examine the refraction of light at a surface between two transparent materials quantitatively

- To understand Snell's law of refraction

- To observe total internal reflection and discover under what circumstances it occurs

- To examine dispersion of light and the formation of rainbows

OVERVIEW

Many optical elements like mirrors and prisms are made of transparent materials such as glass or Lucite with smooth surfaces. As you have observed *qualitatively* in the previous lab, when light is traveling in air and comes upon the surface of a transparent material, some of the light is reflected and some of it is transmitted. To understand how optical elements work, it is important to have a *quantitative* understanding of *reflection* and *refraction* at a surface. You will examine these in Investigation 1.

There is one circumstance in which light traveling in a dense transparent material is only reflected and not transmitted at a surface with a less dense transparent medium on the other side. This phenomenon of *total internal reflection* will be explored in Investigation 2.

Finally, you are well aware from your everyday observations of the separation of white light into its component colors by a prism, and by water droplets to form rainbows. This process of *dispersion* will be explored in Investigation 3.

INVESTIGATION 1: LAWS OF REFLECTION AND REFRACTION

You have observed the light reflected from a surface in the previous lab. Based on these observations, make the following predictions.

Prediction 1-1: Light from a laser is incident as shown on the flat surface of a transparent material. Sketch the direction of the reflected ray, and in words compare it to the incident ray in terms of the following: (1) angle made with the normal to the surface, (2) plane in which it travels, and (3) intensity.

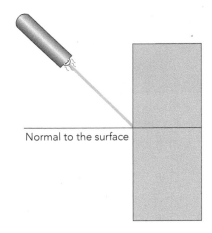

To test your prediction, you will need

- laser

- glass rod

- semicircular transparent chamber filled with a *slightly* cloudy liquid

- blackboard eraser and chalk dust

- sharp pencil

- transparent plastic ruler

- protractor

> **Warning:** The laser beam is intense enough to burn the retina of your eye. Never look directly into a laser or let the laser beam shine directly or reflect into anyone's eye.

Activity 1-1: Law of Reflection

1. Place the chamber on a separate piece of paper, as shown in the diagram that follows. Outline the chamber on the paper with a pencil.

2. Position the laser and glass rod as shown. (The only function of the rod is to spread out the laser beam vertically so that it is more easily seen on the

paper.) The laser beam should be as close to the paper as possible, and should hit the chamber near the center of its flat surface.

3. Mark two points on the paper along the incident beam using a sharp pencil.

4. Mark the point where the laser beam hits the surface.

5. Use the chalk dust to observe the reflected beam. Then *carefully and precisely* mark two points on the paper along the reflected beam using the pencil.

6. Rearrange the laser to change the angle of incidence but have the beam hit the surface at the same point. Again *carefully and precisely* mark the incident and reflected rays with two points for each.

7. Remove the paper, and draw in the incident and reflected rays for both cases, using the ruler. Carefully draw the normal to the surface using the protractor. Save this paper and attach it to these sheets.

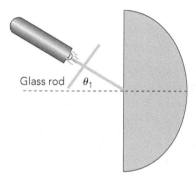

8. Measure the angle of incidence, θ_1, and the angle of reflection, θ_i, for each case. Note that these are conventionally measured from the normal to the surface as shown in the diagram (not from the surface). Record these in Table 2-1, below, along with an estimate of how precisely you can measure these angles given the thickness of the beams, the thickness of the lines, etc.

Table 2-1

	θ_1	θ_i	Estimate of precision
Case 1			
Case 2			

Question 1-1: What do you conclude about the relationship between the angle of incidence and the angle of reflection?

Question 1-2: We can define the *plane of incidence* as the plane defined by the incident beam and the normal to the surface of the transparent chamber. Based on your observations, is the reflected beam also in this plane? Explain how you reached your conclusion.

Question 1-3: The *law of reflection* describes the direction of the reflected beam given the direction of the incident beam. Based on your answers to Questions 1-1 and 1-2, state the law of reflection in words.

Question 1-4: Compare the reflected intensity with the incident intensity. Where did the rest of the light go?

Question 1-5: What about the transmitted light? In what direction does it travel in the liquid in the transparent chamber?

Prediction 1-2: Light from a laser is incident as shown on the flat surface of a transparent material. Sketch the direction of the ray transmitted through the first (left) surface, and in words compare it to the incident ray in terms of the following: (1) angle made with the normal, (2) plane in which it travels, and (3) intensity.

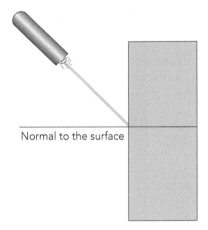

Normal to the surface

To test your predictions, you will need the following in addition to the materials for Activity 1-1:

- computer-based laboratory or graphical analysis software
- *RealTime Physics Light and Optics* experiment configuration files.

Activity 1-2: Snell's Law of Refraction

1. Set up the chamber, laser, and glass rod on a piece of paper as in Activity 1-1. The laser beam should hit near the center of the flat surface.

2. Use the pencil to outline the chamber on the paper.

3. Mark two points on the paper along the incident beam using the pencil and label them with an A.

4. Mark the point where the laser beam hits the surface.

5. Now look at the light transmitted through the flat surface, and mark the point where the ray hits the curved surface. Label this point with an A.

6. Rotate the laser to change the angle of incidence, but make sure that the beam still hits the flat face of the chamber at the same point that you marked previously on the paper.

7. Mark two points on the new incident ray as before, and label them with a B.

8. Again see where the transmitted ray hits the curved surface, and mark that point. Label it with a B.

9. Repeat 6, 7, and 8 until you have six different angles of incidence, including $0°$ (i.e., the laser beam perpendicular to the flat surface, along the normal).

10. Remove the paper and use the ruler to carefully draw in the six incident rays, the six transmitted rays, and the normal. Save this paper and attach it to these sheets.

Question 1-6: Compare the angles of incidence and the angles the transmitted rays make with the normal. If one is always smaller, describe which one.

Comment: The "bending" of light rays as they move from one transparent material (e.g., air) into another (e.g., the fluid in the transparent chamber) is known as *refraction*. The transmitted light is said to be *refracted* at the surface, and the angle measured with the normal is called the *angle of refraction*.

---

11. For each case, measure the angle of incidence, θ_1, and the angle of refraction, θ_2. Enter the values in Table 2-2.

12. Enter the values of $\sin \theta_1$ and $\sin \theta_2$ in the table.

13. Open the experiment file **L02A1-2 (Snell's Law)** to display a table and graph window. Enter the values from Table 2-2 in order from smallest to largest θ_1.

Table 2-2

	θ_1	$\sin \theta_1$	θ_2	$\sin \theta_2$	Estimate of precision
Case A					
Case B					
Case C					
Case D					
Case E					
Case F					

14. Use the **fit routine** in the software to find the relationship between $\sin \theta_1$ and $\sin \theta_2$. Print the graph along with the fit, and attach it to these sheets.

Question 1-7: Based on your analysis, what is the relationship between $\sin \theta_1$ and $\sin \theta_2$? Write this as a mathematical equation.

Comment: Snell's law of refraction states the relationship between θ_1 and θ_2. It says $n_1 \sin \theta_1 = n_2 \sin \theta_2$, where the numbers n_1 and n_2 are quantities called the *indexes of refraction*, which describe the optical properties of the two materials.

Question 1-8: Based on your graph, and given that the index of refraction of air is 1.00, what is the index of refraction of the liquid in the transparent chamber? Be sure to use data from your graph rather than individual values from the table. Show and explain your calculations.

INVESTIGATION 2: TOTAL INTERNAL REFLECTION

Fiber Optics

Today, most of the phone calls in the world travel for at least part of their journey as light signals on fiber optic cables. Because of the high frequencies of light waves compared to microwave and radio waves, these cables are able to carry many more phone messages than metal wires.

But, how is it possible for light to travel along a transparent glass fiber without "leaking" out? In this investigation, you will explore *total internal reflection*, the phenomenon that allows light to travel long distances inside a light fiber.

In Activity 1-2 you collected data for light traveling in air, which has an index of refraction of 1.00 ($n_1 = 1.00$), when it is refracted by water ($n_2 = 1.33$). In other words, $n_2 > n_1$.

Question 2-1: Based on your data for Activity 1-2, for this situation which angle is always greater, θ_2 or θ_1? Explain your answer based on Snell's law, $n_1 \sin \theta_1 = n_2 \sin \theta_2$. (**Note:** Remember that the angles are measured *from the normal, not from the surface.*)

Comment: When it is refracted at the boundary with a medium with a larger index of refraction than the one the light is originally traveling in (what we call *a more optically dense medium*), we say that the light is bent *closer to the normal.*

Prediction 2-1: Suppose that the laser light is originally traveling in the water, and is refracted at the surface with the air. (That is, the light is refracted at the boundary with a *less optically dense medium, $n_1 > n_2$.*) Which angle do you think will now be larger, θ_2 or θ_1? (**Note:** Again, remember that the angles are measured *from the normal not the surface. Hint:* Look at Snell's law.)

To test your predictions, you will need the following:

- laser
- glass rod

- semicircular transparent chamber filled with a *slightly* cloudy liquid
- blackboard eraser and chalk dust
- pencil
- transparent plastic ruler
- protractor

Activity 2-1: Refraction at a Less Dense Medium

1. Set up the laser, glass rod, and transparent chamber on a separate piece of paper as shown in the diagram below. Choose an initial θ_1 of 20–30°. Outline the chamber on the paper with a pencil.

   This time the laser beam enters the chamber through the curved face. In this activity, be sure that the beam always travels along a radius, and, therefore, always strikes the center of the flat surface for every incident angle, θ_1.

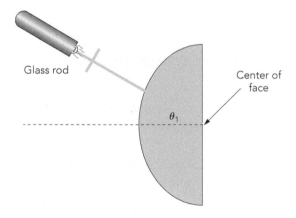

Question 2-2: Why is it important that the laser beam travel along a radius? (*Hint:* In this case, what are the incident and refracted angles at the curved face according to Snell's law? Is the beam bent at the curved surface in this case?)

2. Observe the direction of the refracted beam.

Question 2-3: Qualitatively compare θ_2 to θ_1. Which is larger? Does this agree with Prediction 2-1? Explain.

Prediction 2-2: Suppose that you move the laser and increase the angle of incidence, θ_1. What will happen to θ_2?

Prediction 2-3: Is there an angle of incidence for which no light is transmitted from the water into the air? If so, what is that angle? (*Hint:* Remember that θ_2 cannot be greater than 90°. Calculate the corresponding θ_1, given that $n_1 = 1.33$ and $n_2 = 1.00$.)

3. Increase θ_1 by moving the laser. Remember to be sure that the beam always travels along a radius. Observe the refracted ray as you increase θ_1 slowly up to about 60°.

Question 2-4: Describe what happened to θ_2 as θ_1 was increased.

Question 2-5: What about Prediction 2-2? Is there a value of θ_1 for which there is no transmitted ray—for which all of the light is reflected back into the water?

Activity 2-2: Disappearing Transmitted Ray—A More Quantitative Look

1. With the same setup as in Activity 2-1, use the method of Activity 1-2 to measure θ_1 and the corresponding θ_2 for several different incident angles. Record your values in Table 2-3.

2. Carefully locate and measure the smallest incident angle for which there is no transmitted ray. Record this angle in the table.

Table 2-3

	θ_1	θ_2	Estimate of precision
Case 1			
Case 2			
Case 3			
No transmitted ray		███	

Question 2-6: Does your value for θ_1 when there is no transmitted ray agree with the value you calculated in Prediction 2-3?

Comment: The incident angle for which there is *total internal reflection* is called the *critical angle*. Note that no energy is lost when light is totally internally reflected. Virtually all of the incident intensity is reflected back into the optically denser material.

Question 2-7: Explain how total internal reflection enables the design of light "pipes" or cables, through which light can be transmitted over long distances without leaking out. Make a sketch to illustrate your ideas.

INVESTIGATION 3: REFRACTION, DISPERSION, TOTAL INTERNAL REFLECTION, AND THE RAINBOW

Light from the sun or other white light sources includes all of the colors from violet (shortest wavelength) to red (longest wavelength). The index of refraction for most materials is slightly different for the different colors (wavelengths) of light. This phenomenon is called *dispersion*.

Prediction 3-1: Based on Snell's law, if the index of refraction of a material varies with wavelength, what effect would this have on the refraction of different colors of light?

Prediction 3-2: Suppose that red laser light is incident on a prism—a triangular-shaped slab of glass—as shown below. Sketch and describe the refracted light that comes out the other side of the prism.

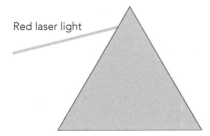

Red laser light

Prediction 3-3: Suppose that white light is incident on a prism as shown below. Sketch and describe the refracted light that comes out the other side of the prism. (*Note*: Because of dispersion, the index of refraction is different for each color of light. Make a guess as to whether the index is largest for red light or for violet light. State your guess, and base your predictions on it.)

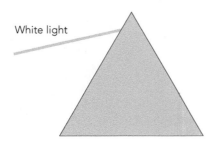

White light

To test your predictions you will need

- laser
- flashlight with a narrow slit
- prism

Activity 3-1: Dispersion of White Light by a Prism

1. Set up the flashlight, slit, and prism as shown in the diagram below.

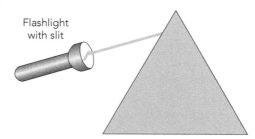

Flashlight
with slit

2. Observe the light coming out from the right side of the prism. Use a piece of paper as a screen.

3. Make a sketch above to the right of the prism of the pattern of the refracted light that you observed. Be sure to label the color(s).

Question 3-1: Compare your observations to Predictions 3-1 and 3-3. Do you see a rainbow of colors? Which color is refracted the most by the prism and which the least?

Question 3-2: Based on this observation, is the index of refraction larger for red or violet light?

Activity 3-2: Dispersion of Laser Light by a Prism

1. Set up the laser and prism as shown in the diagram below.

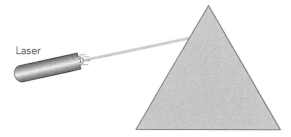

Laser

2. Observe the light coming out from the right side of the prism with a piece of paper—**DO NOT LOOK INTO THE LASER BEAM.** Be sure to move the paper up and down to observe everything.

3. Make a sketch above to the right of the prism of the pattern of the refracted light that you observed. Be sure to label the color(s).

Question 3-3: Compare your observations to Prediction 3-2. Do you see a rainbow of colors? Explain any ways in which your observations are different than for the white light from the flashlight.

If you have time, do the following extension.

Extension 3-3: The Rainbow

Now you are ready to explore how *refraction, dispersion,* and *total internal reflection* in spherical water droplets result in the beautiful rainbows we observe in the sky. You will need

- flashlight with slit
- solid transparent cylinder

1. Set up the flashlight, slit, and transparent cylinder as shown in the diagram below.

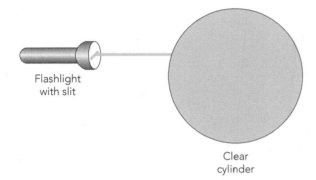

Flashlight
with slit

Clear
cylinder

2. Move a piece of paper all the way around the transparent cylinder, and observe the color(s) of any transmitted light and also the direction in which transmitted light is observed. (You may also want to look into the cylinder with your eye.) (*Hint*: Think about locations of the sun, water vapor droplets (up in the sky) and your eyes when you view a real rainbow.)

3. Sketch your observations on the diagram, indicating clearly the direction in which light leaves the cylinder. Be sure to label the color(s).

4. Sketch the path of the rays inside the cylinder.

Question E3-4: Describe what you observed. Did you see a rainbow? If so, in which direction? (Answer this based on the geometry in the diagram above.)

Question E3-5: Try to explain what you observed based on what you know about *refraction, dispersion,* and *total internal reflection.* Draw a diagram and sketch rays inside the cylinder.

Question E3-6: In the diagram below, the circles represent water droplets in a cloud. The arrows represent rays of light from the sun. Where should you stand to observe the rainbow? Explain based on your observations in this activity.

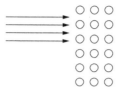

Question E3-7: In this activity you used a transparent *cylinder*, but rain droplets are actually *spherical*. Try to explain why rain droplets arranged as shown in Question E3-6 produce a semicircular *bow* of colors.

Name_____ Date_____ Partners_____

HOMEWORK FOR LAB 2: REFLECTION AND REFRACTION OF LIGHT

1. Light in air from a laser is incident on a slab of glass with angle of incidence 30°, as shown on the right. The index of refraction of the glass is 1.50. Sketch the reflected ray in the diagram, calculate the angle of reflection, and mark the angle of reflection on the diagram.

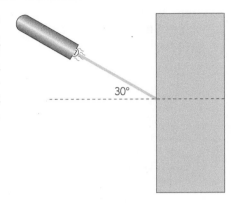

2. Sketch the *refracted* ray in (1) inside the slab on the diagram above. Calculate the angle of refraction, and mark it on the diagram.

3. With the situation illustrated in (1) is it possible for total internal reflection to take place? Explain.

4. Calculate the critical angle for the glass in (1).

5. Laser light traveling inside a slab of glass of index of refraction 1.50 is incident on the surface with an angle of incidence of 60°, as shown on the right.

   a. Sketch the reflected ray (if there is one), calculate the angle of reflection, and mark the angle of reflection on the diagram.

   b. Sketch the refracted ray (if there is one), and mark the angle of refraction on the diagram.

   c. If either the reflected or refracted ray is missing, explain why.

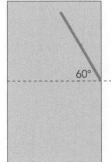

6. White light is incident on a prism as shown. Sketch the light when it leaves the prism, and indicate where the red, green, and violet light will be found. Explain why the transmitted light appears this way instead of white.

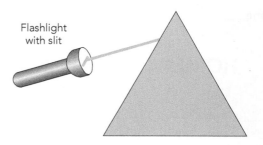

Flashlight with slit

Which color is bent the most, and which is bent the least? Explain why this is true.

Pre-Lab Preparation Sheet for
Lab 3—Geometrical Optics—Lenses

(Due at the beginning of Lab 3)

Directions:
Read over Lab 3 and then answer the following questions about the procedures.

1. What are the two miniature light bulbs used for in Investigation 1?

2. Why are various portions of the lens blocked in Activity 1-1?

3. Why is a very distant light source used in Activity 2-1?

4. What is a *virtual* image? How is it different from a *real* image?

5. How can you tell a diverging lens from a converging one?

LAB 3:
GEOMETRICAL OPTICS—LENSES

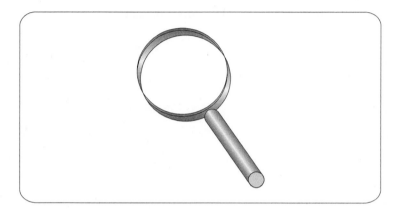

To me the fundamental supposition itself seems impossible, namely, that the waves or vibrations of any fluid can, like the rays of light, be propagated in straight lines, without continual spreading and bending every way into the quiescent medium.

—Isaac Newton, 1704

OBJECTIVES

- To explore how a lens forms images

- To examine how a converging lens focuses light from a very distant object

- To measure the focal length of a converging lens

- To examine how a converging lens forms images of objects that are not very distant

- To examine how a diverging lens forms images

- To measure the focal length of a diverging lens

OVERVIEW

Lenses are essential elements in most optical instruments, from eye glasses to a simple magnifier, to binoculars, optical telescopes, and microscopes. In Investigation 1, you will explore how images are formed by lenses. You will see that for a sharp image to be formed, it is essential that *all* the rays leaving each point on the object are focused by the lens to a corresponding, unique point on the image.

In the rest of this lab, you will look more quantitatively at image formation by lenses. In Investigation 2, you will look at the action of a converging lens in forming an image. You will examine the definition of *focal length,* and see the relationship between object distance, image distance, focal length, and magnification. In Investigation 3, you will examine the images formed by diverging lenses.

INVESTIGATION 1: HOW ARE IMAGES FORMED BY LENSES?

When light is emitted or reflected by an object, each point on the object serves as a source of light. The light from each of these points spreads out in all directions in space. To understand what we see or what image is formed by a lens or mirror, we must first see what happens to the light from each of these point sources. The filament of a light bulb, while not exactly a point source of light, is small enough to be a good approximation.

To simplify matters, in this activity you will focus on the rays coming from only two points of an object.

You will need

- two miniature light bulbs
- power supply
- comb
- cylindrical lens
- green filter

Activity 1-1: A Simple Real Image

Imagine that the arrow drawn in the picture on the following page is lighted. (It is either a source of light, or light is reflecting off of it.) Every point on the arrow sends out rays in all directions. To simulate this, you will put small light bulbs at the head and foot of the arrow. These bulbs will represent two points of the arrow that are emitting light.

1. Tape one of the miniature bulbs to the top and one to the bottom of the arrow, on the next page, so that both bulbs are pointing straight up perpendicular to the paper. (Do not place the cylindrical lens on the paper yet.)

2. Connect bulb 1 to the power supply and turn it on. Then disconnect bulb 1 and connect bulb 2.

Question 1-1: Describe what the light from bulb 1 does. Describe what the light from bulb 2 does.

3. Place the cylindrical lens in the semicircular outline. Sketch on the diagram the beam of light that comes out of the lens with only bulb 2 turned on.

4. Place the comb midway between the bulb and the lens, parallel to the flat face of the lens, so that the beam is divided up into rays.

Question 1-2: Describe what the lens does to the rays from bulb 2. What is their direction when they leave the bulb (diverging, parallel, or converging)? What is their direction when they hit the front surface of the lens? What is their direction when they leave the lens?

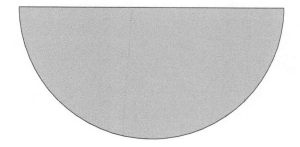

5. Remove the comb, and place an X at the point where the image of bulb 2 is formed (where the rays from bulb 2 are converged to the smallest point by the lens).

Question 1-3: Are rays from bulb 2 hitting only part of the front surface of the lens or all of the front surface?

6. Repeat (5) for bulb 1.

7. Now turn both bulbs on at the same time. Put the green filter in front of bulb 2 so that you can distinguish the light from the two bulbs. Draw an arrow on the paper at the location where the image would be formed by the lens.

Question 1-4: Would the image of the arrow be upright or inverted? How do you know?

Question 1-5: Is the image of the arrow enlarged or reduced in size compared to the arrow itself?

Prediction 1-1: Suppose that you moved the lens further away from the arrow. How would the image be changed?

Prediction 1-2: Suppose that you moved the lens closer to the arrow. How would the image be changed?

Test your predictions.

8. Move the lens further away from the bulbs.

Question 1-6: Describe what happens to the image. Is it now larger or smaller? How do you know?

9. Move the lens closer to the bulbs.

Question 1-7: Describe what happens to the image. Is it now larger or smaller? How do you know?

Question 1-8: Is there a position of the lens closer to the bulbs beyond which an image is no longer formed? Experiment and explain.

Prediction 1-3: Suppose that you covered half of the side of the lens facing the arrow with a card, i.e., covered the top half of the lens as seen in the diagram above. How would the image be changed? Would the whole image of the arrow still be formed?

10. Move the lens back on the outline. Block the top half of the lens (as seen in the diagram above) with a card.

Question 1-9: Carefully describe what happens to the image. Explain your observations based on what happens to rays from each of the bulbs that hit the unblocked half of the lens.

Prediction 1-4: Suppose that you covered the center of the lens facing the arrow with a piece of paper. How would the image be changed? Would the whole image of the arrow still be formed?

11. Block the center of the lens with a piece of paper.

Question 1-10: Carefully describe what happens to the image. Explain your observations based on what happens to rays from each of the bulbs that hit the unblocked portions of the lens.

Prediction 1-5: Suppose that you covered the top half of the arrow with a piece of paper. How would the image be changed? Would the whole image of the arrow still be formed?

12. Block the top bulb (bulb 1) with a card.

Question 1-11: Carefully describe what happens to the image. Explain your observations based on what happens to rays from each of the bulbs.

Prediction 1-6: Suppose that you removed the lens. How would the image be changed? Would the whole image of the arrow still be formed?

13. Remove the lens.

Question 1-12:   Carefully describe what happens to the image.

Question 1-13:   Carefully describe the function of a lens in forming an image. Describe what the lens does to the light coming from each point on the object.

INVESTIGATION 2: CONVERGING LENSES

To examine the image formed by a converging lens, you will need

- two different double convex lenses
- screen
- optical rail and holders
- clear bright light bulb and socket (distant light source)
- arrow-shaped object light source
- transparent metric ruler
- meter stick

Activity 2-1: Image of a Distant Object Formed by a Converging Lens—Focal Length

To observe the image of a distant object formed by a double convex lens (the lens with both faces bulging out at the center), first mount the lens in the lens holder on the rail. Also mount the screen on a holder on the rail.

1.   For the object, use a light source that is very distant from your setup (across the room). Move the screen until a sharp image of the filament of the light bulb is formed on the screen. Then move the screen a few centimeters closer to the lens and further away from the lens and observe what happens to the image.

Question 2-1:   Describe your observations. Would you conclude that this lens is able to form a sharp image of a distant object? Explain.

Question 2-2:   Describe what happened to the image when you moved the screen a little closer and a little further away from the lens. Does there seem to be only one position of the screen where the image is really sharp?

> **Comment:**   Light can be represented by *wave fronts* or by *rays*. As you have seen in Lab 1, for light from a very distant point source the wave fronts are nearly parallel planes, and the rays are parallel lines perpendicular to these wave fronts.

As you can see in Figure 3-1, a converging lens bends these rays so that they converge and intersect at the point where the image is formed, called the image point.

The distance from an object to the lens is called the *object distance*. The distance from the lens to the location of the image is called the *image distance*. In this special case in which the object is *very distant* from the lens, the image point is called the *focal point,* and the image distance is called the *focal length*. The focal length is a number that characterizes how effective a lens is in focusing light rays.

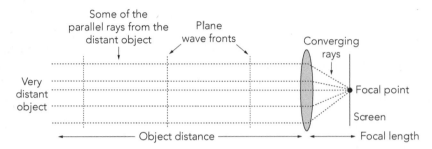

Figure 3-1: Definition of the *focal point* and *focal length* of a converging lens.

2. Measure the focal length of the lens. Move the screen back so that the image is sharp once again. Read the position of the lens and the position of the screen as accurately as you can on the scale, and determine the distance from the lens to the image. Repeat five times and take the average value as the focal length.

Position of lens A: _____ cm

Positions of screen: _____ _____ _____ _____ _____ cm

Distance lens to screen: _____ _____ _____ _____ _____ cm

Average focal length of the lens: _____ cm

3. Repeat, and measure the focal length of a different double convex lens.

Position of lens B: _____ cm

Positions of screen: _____ _____ _____ _____ _____ cm

Distance lens to screen: _____ _____ _____ _____ _____ cm

Average focal length of the lens: _____ cm

Question 2-3:   How well could you measure the focal length of each lens, i.e., how much variation was there in your measurements?

Question 2-4:   Do the two double convex lenses have different focal lengths?

Question 2-5:   Observe the two lenses. Does one appear to be more curved than the other, i.e., does one seem to bulge more at the center? Which lens has the smaller focal length? How does the curvature of the surfaces of a double convex lens seem to relate to the ability of the lens to focus parallel rays of light?

The focal point is the image point for objects that are very far from the lens. Suppose that the object is closer to the lens. You will explore this in the next activity. You will need the same equipment as for Activity 2-1.

Activity 2-2: Image Formation by a Converging Lens

Here you will position the arrow-shaped object light source at different distances from the lens, and you will observe the image on the screen. You will examine what happens to the location and size of the image as the object is moved relative to the lens.

1. Place one of the converging (double convex) lenses and the screen on the rail. Also mount the object light source on a holder.

2. Position the object light source so that the arrow is 50 cm from the lens, on the opposite side of the lens from the screen. (The light source may be off the end of the rail.)

3. Measure the length of the arrow. This is the object size (length).

    Object size: _____ cm

Prediction 2-1: With the object distance now only 50 cm instead of very large, will the image distance be larger, the same as, or smaller than the focal length measured in Activity 2-1? Explain how you reached your conclusion. A diagram might be helpful.

> **Comment:** For an object distance that is *not* very large, *the image location is no longer referred to as the focal point and the image distance is no longer referred to as the focal length.* These special names are reserved for very large (infinite) object distance and incident parallel rays (plane waves), as in Activity 2-1.

4. Find the location of the image. Move the screen until a sharp image is formed. Record the positions of the lens, object, and screen in the first line of Table 3-1. Calculate the object distance and the image distance. Also record the size (length) of the image, calculate the *magnification* (ratio of image size to object size), and state whether the image is *upright* or *inverted*. (Don't worry about the other rows in the table for now.)

Question 2-6: Was your prediction correct? If not, try to explain your observations.

5. View the image with your eyes looking through the lens. Remove the screen. Place your head behind the location where the screen was and look back through the lens.

Table 3-1

Position of lens (cm)	Object		Image		Image size (cm)	Magnification (image size/ object size)	Upright or inverted?
	Position (cm)	Distance (cm)	Position (cm)	Distance (cm)			
		50.0					

Question 2-7: Do you see the same image that was on the screen before you re-moved it? Describe the image that you observe. Is it upright or inverted? Bigger or smaller than the object? On which side of the lens does the image appear to be—the object side or the side where your eye is? (*Hint:* Try to pinch the image.)

6. Replace the screen and try other object distances. Moving the object source 10 cm closer to the lens each time, move the screen to find the new image location for a number of object distances. Record the lens, object, and image positions, the image size, and other information requested in Table 3-1. Cal-culate the object and image distances.

   When the object distance is about 5 cm larger than the *focal length* (mea-sured in Activity 2-1), begin to move the object in just 1 cm steps. (At some point, you may need to move the screen beyond the end of the rail, use a larger screen, or observe the image on the wall across the room.)

7. Is there a closest object distance beyond which no image is formed on the screen? Locate this object distance as best you can, record it in the table, and indicate that no image was formed.

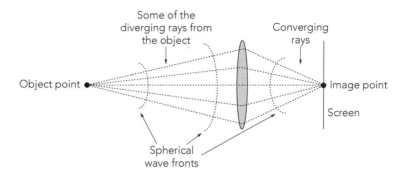

Figure 3-2: The action of a converging lens to form a *real* image.

Question 2-8: What happens to the image distance when the object distance is made smaller and smaller?

Question 2-9: What happens to the image size when the object distance is made smaller and smaller? What happens to the magnification?

Question 2-10: Compare the minimum object distance for which an image is formed on the screen with the focal length measured in Activity 2-1. Are they the same? Explain?

Comment: When the object isn't very distant from the lens, the rays diverge from the object, and the wave fronts are spherical. A converging lens bends these rays. As you can see in Figure 3-2, if it is able to bend them enough, they will converge to a point, and an image will form on the screen.

Question 2-11: How are focusing *power* and *focal length* of a lens related?

Comment: All of the images observed above are said to be *real*. That is, they are formed when rays from points on the object are bent by the lens and converge to points on the screen.

You also observed in this activity that if the object is closer to a converging lens than the focal point, no real image can be formed on a screen. *This does not mean there is no image!* In the following activity, you will explore this further and look for a different kind of image.

Activity 2-3: Looking for a *Virtual* Image

1. Set up the object about 5 cm closer to the lens than the focal point. Look through the lens from the side opposite the object toward the object. Observe the image you see very carefully.

Question 2-12: Describe the image. Is it *upright* or *inverted?* Is it larger or smaller than the object? (It will be helpful to remove and reinsert the lens to check this.)

Question 2-13: Can you observe this image on the screen? (Try to insert the screen if you need to.) Explain.

> **Comment:** This image is said to be *virtual* rather than *real,* since it isn't formed by the real convergence of light rays at an object point. When the object is closer to the lens than the focal point, as in Figure 3-3, the lens isn't powerful enough to focus the diverging rays to a point. Instead the rays continue to diverge. When you look back into the lens, these rays *appear to be* diverging from a point on the same side of the lens as the object point. This point is called the *virtual image point.*

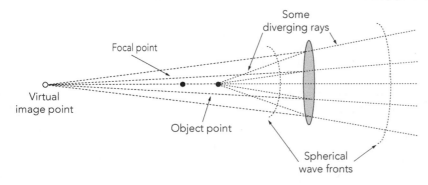

Figure 3-3: The action of a converging lens to form a *virtual* image.

2. Observe the image for another object distance. Move the object still closer to the lens. (You may need to move your head further away from the lens to see the image clearly.)

Question 2-14: Describe the image. Is it *upright* or *inverted?* Observe very carefully. Is it larger or smaller than the object? Is it larger or smaller than the image you observed in (1)? (It will be helpful to remove the lens and also move the object back to its position in (1) to make these observations.)

Question 2-15: If the object is closer to a converging lens than the focal point, is it possible for the image to be smaller than the object? (If necessary, make some more observations to check your answer.)

Question 2-16: What simple optical device uses a converging lens in this way?

If you have more time, work on the extensions in Investigation 3.

INVESTIGATION 3: DIVERGING LENSES

In Investigation 2, you looked at image formation by a converging lens. You saw that when the object was further from the lens than the focal point, the image formed was real and inverted. When the object was closer to the lens than the focal point, the image formed was virtual and upright. In this investigation you will examine the images formed by a *diverging* lens.

You will need

- double concave lens
- screen
- optical rail and holders
- clear light bulb and socket (distant light source)
- arrow-shaped object light source

Extension 3-1: Can a Real Image Be Formed by a Diverging Lens?

1. Set up the double concave lens (the one with both faces curved inward) and screen on the rail. Try to locate the image of the filament of the light bulb (across the room) on the screen.

Question E3-1: Are you able to locate an image on the screen? Can you locate an image anywhere? Describe how your observations compare to those for the converging lens in Activity 2-1.

Question E3-2: From your observations, what is different about the effect of converging and diverging lenses on parallel rays of light?

Comment: As you can see in Figure 3-4, a diverging lens (such as a double concave lens) bends parallel rays so that they appear to diverge from a point after they pass through the lens. This point is a *virtual image point*. It is the focal point of the lens. As for a converging lens, the focal length is defined as the distance from the focal point to the lens.

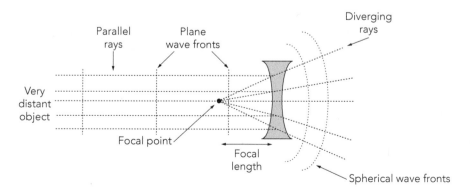

Figure 3-4: The action of a diverging lens.

Table 3-2

| Position of lens (cm) | Object | | Image | | Image size (cm) | Upright or inverted? |
	Position (cm)	Distance (cm)	Position (cm)	Distance (cm)		
		50.0				

2. Mount the arrow-shaped object light source on a holder, and mount the diverging lens and the screen on the rail.

3. Position the object light source so that the arrow is 50 cm from the lens, on the opposite side of the lens from the screen. (The light source will be off the end of the rail.)

4. For different distances of the object source from the lens, try to locate the image on the screen. If you can locate the image at same distances, record these data in Table 3-2. Also record in the table object distances for which you cannot locate an image on the screen.

Question E3-3: What do you conclude about the formation of real images by a double concave (diverging) lens?

Extension 3-2: Formation of Virtual Images by a Diverging Lens

1. Remove the screen from the rail. Look through the diverging lens from the side opposite the object toward the object.

Question E3-4: Describe the image that you observe. Is it upright or inverted? Is it larger or smaller than the object?

Question E3-5: Is this a *real* or *virtual* image? Explain.

2. Try a different object distance. Move the object much closer to the lens and again carefully observe the image through the lens.

Question E3-6:   Describe the image. Is it upright or inverted? Is it larger or smaller than the object? Is it larger or smaller than the image you observed in (1)? (Move the object further away, then closer, to observe this.)

Question E3-7:   Is it possible for a diverging lens to form a *real* image? Is it possible for a diverging lens to form an *inverted* image?

Question E3-8:   Is it possible for a diverging lens to form an image that is larger than the object? (If necessary, make some more observations to check your answer.) Explain.

Question E3-9:   What kinds of images can be formed by a diverging lens?

Comment: As you can see in Figure 3-4, the focal point for a diverging lens is a *virtual* focal point. It is the point from which parallel rays seem to diverge after passing through the lens. The focal length is also defined in the diagram. It can be measured using light from a distant source, as was done for a converging lens in Activity 2-1, but the measurement is more difficult.

In addition to the equipment you have been using, to measure the focal length of a diverging lens. you will need

- transparent metric ruler

Extension 3-3: Focal Length of a Diverging Lens

1.   Set up the diverging lens and screen on the rail, as in Extension 3-1. Also set up the distant object source (across the room).

2.   Position the screen about 20 cm behind the lens. Measure this distance carefully.

   Distance from lens to screen:_____ cm

3.   Now, instead of looking for an image of the distant source, look for the circle of illumination from the distant source on the screen. Measure the diameter of this circle and the diameter of the lens.

   Diameter of circle of illumination:_____ cm

   Diameter of lens:_____ cm

Question E3-10:   Look at Figure 3-4 and explain the origin of this circle of illumination.

4. Use your measurements to determine the focal length of the diverging lens. Refer to Figure 3-4. (The diagram that follows may be helpful.) Show your calculation below. (Note that the focal length for a virtual focal point is a *negative* number.)

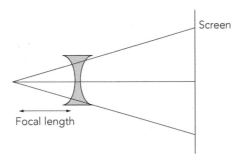

Focal length:_____ cm

Question E3-11: How accurately do you think you can measure the focal length by this method? What are some of the problems that make this measurement difficult?

Prediction E3-1: Suppose that the double concave (diverging) lens and a double convex (converging) lens are attached together face to face to form a single lens. How would the focal length of this lens combination compare to the focal length of the converging lens alone measured in Activity 2-1? Explain how you arrived at your prediction.

To test your prediction you will need the equipment above plus

- double convex lens from Investigation 2
- tape to attach the two lenses

Extension 3-4: Focal Length of a Lens Combination

1. Attach the two lenses to form a single lens. Put the lenses face to face and tape their edges to hold them together.

2. Determine the focal length of this lens combination. Mount the lens combination and the screen on the rail and determine the focal length by finding the image of the very distant object, as in Activity 2-1.

3. Make five independent determinations of the image distance by moving the screen so that the image is first out of focus and then back in focus. Record your data below.

Location of lens combination (center of the two lenses): _____ cm

Locations of screen: _____ _____ _____ _____ _____ cm

Distance from screen
to lens combination: ———— ———— ———— ———— ———— cm

Focal length of lens combination: ———— cm

Question E3-12: Was your prediction correct? Explain your observations in terms of the way the two lenses converge or diverge parallel light rays.

Comment: When two lenses of focal lengths f_1 and f_2 are combined, their focal lengths add by the following equation:

$$\frac{1}{f_C} = \frac{1}{f_1} + \frac{1}{f_2}$$

where f_C is the focal length of the lens combination.

---

4. Check whether the value for f_C calculated from this equation agrees with the value you just measured. Use your values for the focal length of the double convex (converging) lens (f_1) from Activity 2-1 and the double concave (diverging) lens (f_2) from Extension 3-3. (Remember that the latter is negative.) Show your calculations.

    Calculated focal length of lens combination: ———— cm.

Question E3-13: Do your measured and calculated values for the focal length of the lens combination agree? Can you account for any differences based on the possible inaccuracies in measuring the focal lengths of the two lenses?

HOMEWORK FOR LAB 3: GEOMETRICAL OPTICS: LENSES

1. A camera has a lens (or combination of lenses) like the converging lens in this lab that focuses light from objects forming real images on a piece of film (like the screen in this lab). An enlarger shines light through a negative, and uses a lens to project a real image of the picture on the negative onto the platform where the photographic paper is placed. Explain how each of the following will affect your photographs.

   a. Half of the lens on your camera is covered by a piece of paper.

   b. The negative is placed in the enlarger with half of it covered by a piece of tape on the inside.

   c. Half of the lens on the enlarger is covered by a piece of paper.

   d. The camera lens is replaced by a diverging lens with the same focal length.

2. You have a converging lens of focal length 20 cm. Answer the following based on your observations in this lab.

   a. For what range of object distances will the image be larger than the object?

   b. For what range of object distances will the image be smaller than the object?

   c. For what range of object distances will the image be *upright?*

   d. For what range of object distances will the image be *inverted?*

 e. For what range of object distances will the image be *real?*

 f. For what range of object distances will the image be *virtual?*

3. You have a diverging lens of focal length −20 cm. Answer the following based on your observations in this lab.

 a. For what range of object distances will the image be larger than the object?

 b. For what range of object distances will the image be smaller than the object?

 c. For what range of object distances will the image be *upright?*

 d. For what range of object distances will the image be *inverted?*

 e. For what range of object distances will the image be *real?*

 f. For what range of object distances will the image be *virtual?*

Name_____   Date_____

PRE-LAB PREPARATION SHEET FOR
LAB 4—GEOMETRICAL OPTICS—MIRRORS

(Due at the beginning of Lab 4)

Directions:
Read over Lab 4 and then answer the following questions about the procedures.

1. Define *plane, concave,* and *convex.*

2. What is the purpose of the two miniature light bulbs in Investigation 1?

3. Define *real* and *virtual* as used to describe images.

4. What is your Prediction 1-5?

5. Define *focal point* and *focal length.*

LAB 4:
GEOMETRICAL OPTICS—MIRRORS

There is no mystery in a looking glass until someone looks into it. Then, though it remains the same glass, it presents a different face to each man who holds it in front of him.

—Harold C. Goddard, 1951

OBJECTIVES

- To explore how a mirror forms images
- To examine the differences between concave, plane, and convex mirrors and how they affect light rays
- To examine image formation by a concave mirror
- To examine image formation by a convex mirror
- To examine image formation by a plane mirror

OVERVIEW

Mirrors are common optical elements in our lives. From the moment we wake up in the morning and stare at our faces in the bathroom mirror, we are depending on the properties of mirrors to form images. The rearview mirrors in an automobile, the reflectors in the headlights, and that little mirror your dentist sticks in your mouth to view your teeth are other examples. Mirrors are also found in important scientific devices, with telescopes being the most well-known example. The mirror in an optical telescope can be as large as 5 meters in diameter!

The mirrors in the examples given above are not all the same. Your bathroom mirror has a flat surface, as do the driver side and interior rearview mirrors in a car. These are called *plane* mirrors. But the passenger side rearview mirror is *convex* (bulging outward slightly at the center), and the mirror in a telescope is *concave* (bulging slightly inward at the center). We will explore all three types of mirrors.

Image formation by mirrors is another example of geometrical optics, since mirrors are usually much larger in diameter than the wavelengths of visible light. Just as for lenses, you will see that for a sharp image to be formed, it is essential that *all* the rays leaving each point on the object are focused by the mirror to a corresponding, unique point on the image.

In Investigation 1 you will compare the focusing effects of plane, concave, and convex mirrors. Then, in Investigation 2, you will explore the formation of images by a concave mirror. Finally, in Investigation 3, you will explore how plane and convex mirrors form images.

INVESTIGATION 1: HOW ARE IMAGES FORMED BY MIRRORS?

As we saw in Lab 3, when light is emitted or reflected by an object, each point on the object serves as a source of light that spreads out in all directions in space. Just like in Lab 3, we will first use small light bulbs as point sources and examine what different mirrors do to the light rays that diverge from each point on an object.

Once again, to simplify matters, in this activity you will focus on the rays coming from only two points of an object.

You will need
- two miniature light bulbs
- power supply
- comb
- concave cylindrical mirror
- plane (flat) mirror
- convex cylindrical mirror
- green filter
- transparent ruler

Activity 1-1: Image Formation by a Concave Mirror

Imagine that the arrow drawn below is lighted. (It is either a source of light or light is reflecting off of it.) Every point on the arrow sends out rays in all directions. To simulate this, as in Lab 3 you will put small light bulbs at the head and foot of the arrow. These bulbs will represent points of the arrow that are emitting light.

1. Tape the wires from one of the miniature bulbs to the top and the wires from the other bulb to the bottom of the arrow on the following page, so that both bulbs are pointing straight up perpendicular to the paper. (Do not place the mirror on the paper yet.)

2. Connect bulb 1 to the power supply and turn it on. Observe what the light from bulb 1 does. Then disconnect bulb 1 and connect bulb 2. Observe what the light from bulb 2 does.

3. Place the concave mirror on the outline on the diagram. Sketch on the diagram the beam of light that comes out of the mirror with only bulb 2 turned on.

4. Place the comb midway between the bulb and the mirror, parallel to the arrow, so that the beam is divided up into rays.

Question 1-1: Describe what the mirror does to the rays from bulb 2. What is their direction when they leave the bulb (diverging, parallel, or converging), and what is their direction when they leave the mirror?

5. Place an X at the point where the image of bulb 2 is formed (where the rays from bulb 2 are converged by the mirror to the smallest point).

Question 1-2: Are rays from bulb 2 hitting only part of the front surface of the mirror or all of the front surface?

6. Repeat (5) for bulb 1.

7. Now turn both bulbs on at the same time. Put the green filter in front of bulb 2 so that you can distinguish the light from the two bulbs.

8. Draw an arrow on the paper at the location where the image is formed by the mirror.

Question 1-3: Is the image of the arrow upright or inverted? How do you know?

Question 1-4: Is the image of the arrow enlarged or reduced in size compared to the arrow itself?

Question 1-5: Is the image of the arrow *real* or *virtual?* Explain.

Prediction 1-1: Suppose that you moved the mirror *a little* further away from the arrow. How would the image be changed?

Prediction 1-2: Suppose that you moved the mirror *a little* closer to the arrow. How would the image be changed?

Test your predictions.

9. Move the mirror *a little* further away from the bulbs.

Question 1-6: Describe what happens to the image. Is it now larger or smaller? How do you know?

10. Move the mirror *a little* closer to the bulbs.

Question 1-7: Describe what happens to the image. Is it now larger or smaller? How do you know?

Question 1-8: Is there a position of the mirror to the left of the outline beyond which an image is no longer formed? Experiment and explain.

Prediction 1-3: Suppose that you covered the top half of the side of the mirror facing the arrow with a card, How would the image be changed? Would the whole image of the arrow still be formed?

11. Move the mirror back onto the outline. Block the top half of the mirror (top or bottom as seen in the diagram above) with a card.

Question 1-9: Carefully describe what happens to the image. Explain your observations based on what happens to rays from each of the bulbs that hit the unblocked half of the mirror.

Prediction 1-4: Suppose that you cover the top half of the *arrow* with a card. How would the image be changed? Would the whole image of the arrow still be formed?

12. Block the top bulb with a card.

Question 1-10: Carefully describe what happens to the image. Explain your observations based on what happens to rays from each of the bulbs.

Prediction 1-5: Suppose that you removed the mirror. How would the image be changed? Would the whole image of the arrow still be formed?

13. Remove the mirror.

Question 1-11: Carefully describe what happens to the image.

Question 1-12: Carefully describe the function of a mirror in forming an image. Describe what the mirror does to the light coming from each point on the object.

Prediction 1-6: If you replaced the concave mirror with a plane (flat) one, would your observations above be changed? Would an image still be formed? If yes, would its location and size be the same or different?

Now test your predictions.

Activity 1-2: Image Formation with a Plane Mirror

1. Place the bulbs on the head and tail of the arrow on the diagram that follows as you did in Activity 1-1, and replace the concave mirror with the plane one.

2. Turn on light bulb 2 and observe the light reflected from the mirror. Carefully sketch what you see in the space above.

Question 1-13: Compare the light reflected from the plane mirror to your observations in Activity 1-1 with the concave mirror. Is the light focused to a *real* image point by the plane mirror?

3. Now turn off bulb 2 and turn on bulb 1. Again carefully sketch what you see in the space above.

Question 1-14: What is different and what is the same about the light from bulb 1 as compared to bulb 2. Is the light from bulb 1 focused to a *real* image point by the plane mirror?

4. Move the plane mirror further away and then closer to the bulbs and observe the reflected light.

Question 1-15: Based on your observations, can a plane mirror form a *real* image? A *virtual* image?

Question 1-16: If a *virtual* image were formed, where would it be located?

Prediction 1-7: If you replaced the plane mirror with a convex one, would your observations be changed from those in Activity 1-1 with the concave mirror? Would an image still be formed? If yes, would its location and size be the same or different?

Prediction 1-8: Would your observations with a convex mirror differ from those with the plane mirror? If yes, how would they differ?

Test your predictions.

Activity 1-3: Image Formation with a Convex Mirror

1. Place the bulbs on the head and tail of the arrow on the diagram that follows, and replace the plane mirror with the convex one.

2. Turn on bulb 2 and observe the light reflected from the mirror. Carefully sketch what you see in the space above.

Question 1-17: Compare the light reflected from the convex mirror to your observations in Activity 1-1 with the concave mirror. Is the light focused to a *real* image point by the convex mirror?

3. Now turn off bulb 2 and turn on bulb 1. Again carefully sketch what you see in the space above.

Question 1-18: What is different and what is the same about the light from bulb 1 as compared to bulb 2. Is the light from bulb 1 focused to a *real* image point by the convex mirror?

4. Move the convex mirror further away and then closer to the bulbs and observe the reflected light.

Question 1-19: Based on your observations, can a convex mirror form a *real* image? A *virtual* image?

Question 1-20: If a *virtual* image were formed, where would it be located?

Question 1-21: Compare your observations with the convex mirror to those in Activity 1-2 with the plane mirror. If a *virtual* image were formed with the convex mirror, how would it differ from the *virtual* image formed with the plane mirror?

INVESTIGATION 2: IMAGE FORMATION WITH A CONCAVE MIRROR

You have already observed in Investigation 1 what concave, plane, and convex mirrors do to the diverging rays from a point source. Only a concave mirror is able to reflect rays so that they converge to a *real* image point. In this activity, you will explore more quantitatively how this occurs by tracing individual rays from object points to image points.

You will need

- concave cylindrical mirror
- miniature laser

- glass rod
- pencil
- transparent ruler

Activity 2-1: Focal Length of a Concave Mirror

Recall that for a lens the *focal point* is defined as the point where light from a distance source is focused, and the *focal length* is the distance from the focal point to the lens. This definition is the same for a mirror.

Also recall that light from a distant source is represented by parallel rays. To find the focal point and focal length of the concave mirror, we will use a laser to shine two parallel rays onto it, and see where they are reflected.

1. Position the concave mirror on the diagram that follows, and align it with the picture of the mirror. Be sure to keep the mirror aligned in this way throughout this activity.

2. Align the laser and glass rod so that the beam exactly follows the bottom ray on the diagram.

3. Locate the reflected ray, and mark several points along its path. Remove the laser, and use the ruler and pencil to draw the reflected ray on the diagram.

4. Repeat this procedure for the top ray.

5. Mark and label the focal point on the diagram, and measure the focal length.

    Focal length: _____ cm

Question 2-1: Describe your observations. Would you conclude that this mirror would be able to form a sharp image of a distant object? Explain.

Question 2-2: How would you make different mirror with a longer focal length? Shorter focal length? Is the focal length related to the radius of the cylinder of which the mirror surface is a part? How?

Comment: The actual relationship is $f = R/2$, where f is the focal length of the mirror and R is its radius of curvature.

Question 2-3: Based on your observations, does this relationship seem reasonable? Explain.

Activity 2-2: Image Formation by a Concave Mirror

Remember that rays diverge from points on an object. In Activity 1-1 you saw that a concave mirror can focus these diverging rays to form real image points. In this activity you will explore this more quantitatively. You will use the same equipment as in the previous activity.

In the diagram that follows, the arrow is the object for the concave mirror. While you know that the mirror must focus *all the rays* from each object point to a corresponding image point, you will look at just two of the rays from the head of the arrow and two of the rays from the tail.

1. Position the concave mirror on the diagram that follows. Be sure that the mirror remains in this position throughout this activity.

2. Now align the laser and glass rod as shown so that the beam is exactly aligned with the solid, horizontal ray drawn through the tail of the arrow.

3. Locate the reflected ray, and mark several points along its path. Remove the laser, and use the ruler and pencil to draw the reflected ray on the diagram.

4. Repeat this procedure, this time with the laser beam aligned with the dotted, diagonal ray also through the tail of the arrow. Use the ruler and pencil to draw the reflected ray on the diagram.

5. Identify and label the image point of the tail of the arrow.

6. Now repeat this procedure for two rays through the head of the arrow. Draw in the two incident rays and the two reflected rays, and identify and label the image point of the head of the arrow.

7. Draw in the image of the arrow.

8. Measure the distance from the mirror to the image (the *image distance*):

_____ cm

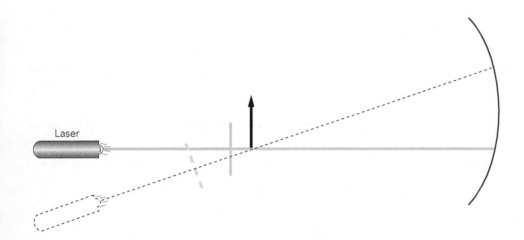

Question 2-4: Explain how you know where the image is formed. Is the image in front of the mirror or behind the mirror?

Question 2-5: Is the image real or virtual? How do you know?

Question 2-6: Is the image upright or inverted? How do you know?

Question 2-7: Is the image enlarged or diminished in size? How do you know?

Question 2-8: Compare the measured image distance for an object that is not very far from the mirror to the *focal length* measured in Activity 2-1. Are these distances the same? State the definition of the *focal length*, and explain if you expect them to be the same.

Prediction 2-1: If you move the object (the arrow) closer to the mirror, are there positions where a *real* image will no longer be formed?

Test your prediction.

Activity 2-3: Virtual Image with a Concave Mirror

1. Set up the concave mirror on the diagram that follows, so that the distance of the arrow from the mirror is about half the focal length measured in Activity 2-1.

2. Repeat the procedures in Activity 2-2 for two rays through the tail of the arrow.

Question 2-9: Do the two rays intersect with each other in front of the mirror? Explain.

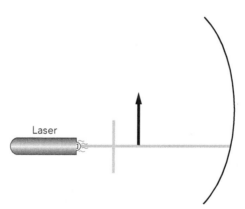

Comment: Recall that when two rays from the same point on the object are bent by an optical element so that they seem to diverge from a point, that point is a *virtual* image point.

---

3. Locate the point from which the two rays seem to diverge, and label it as the *virtual* image point.

Question 2-10: Is the image point located in front of the mirror or behind the mirror?

4. Repeat the procedures above with two rays through the head of the arrow to determine the location of the image point of the head. Label this point.

5. Draw in the image of the arrow.

Question 2-11: Is the image real or virtual? How do you know?

Question 2-12: Is the image upright or inverted? How do you know?

Question 2-13: Is the image enlarged or diminished in size? How do you know?

Question 2-14: Based on your observations here and in Activity 2-2, where should you place an object to get a *real* image formed by a concave mirror? Where should you place an object to get a *virtual* image formed by a concave mirror? Explain.

INVESTIGATION 3: IMAGES WITH CONVEX AND PLANE MIRRORS

You have already made some observations of image formation with plane and convex mirrors in Investigation 1. Based on these observations, make the following prediction.

Prediction 3-1: Can a real image be formed by a convex mirror? Explain based on your observations in Investigation 1.

To test your prediction, you will need the following:

* convex cylindrical mirror
* miniature laser
* glass rod
* pencil
* transparent ruler

Activity 3-1: Image Formation with a Convex Mirror

1. Set up the convex mirror on the diagram that follows. Be sure not to move it from this position throughout this activity.

2. Repeat the procedures in Activity 2-3 for two rays through the tail of the arrow.

3. Locate the real or virtual image point, and label it.

Question 3-1: Do the two rays intersect with each other in front of the mirror? Explain.

Question 3-2: Is the image point located in front of the mirror or behind the mirror?

4. Repeat the procedures above with two rays through the head of the arrow to determine the location of the image point of the head. Label this point.

5. Draw in the image of the arrow.

Question 3-3: Is the image real or virtual? How do you know?

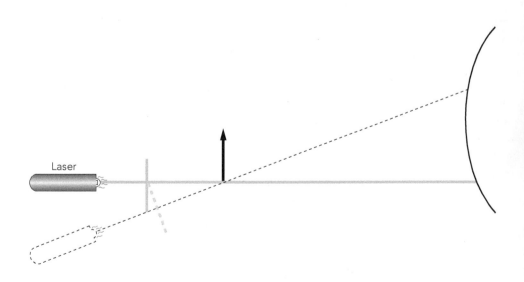

Question 3-4: Is the image upright or inverted? How do you know?

Question 3-5: Is the image enlarged or diminished in size? How do you know?

Question 3-6: Based on your observations here and in Investigation 1, is it possible for a convex mirror to produce a *real* image? Explain.

If you have time, do the following extensions to explore image formation by mirrors further.

Extension 3-2: Image Formation with a Plane Mirror

Prediction E3-2: Can a real image be formed by a plane mirror? Explain based on your observations in Investigation 1.

To test your prediction you will need the equipment above and

- plane mirror

1. Set up the plane mirror on the diagram below. Be sure not to move it from this position throughout this activity.

2. Repeat the procedures in Activity 2-3 for two rays through the tail of the arrow.

3. Locate the real or virtual image point, and label it.

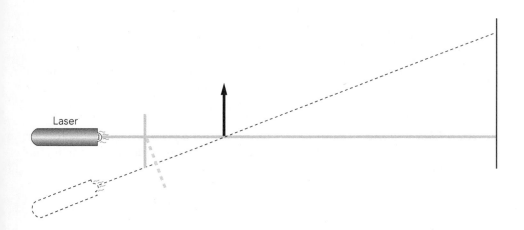

Question E3-7: Do the two rays intersect with each other in front of the mirror? Explain.

Question E3-8: Is the image point located in front of the mirror or behind the mirror?

4. Repeat the procedures above with two rays through the head of the arrow to determine the location of the image point of the head. Label this point.

5. Draw in the image of the arrow.

Question E3-9: Is the image real or virtual? How do you know?

Question E3-10: Is the image upright or inverted? How do you know?

Question E3-11: Is the image enlarged or diminished in size? How do you know?

Question E3-12: Based on your observations here and in Investigation 1, is it possible for a plane mirror to produce a *real* image? Explain.

Extension 3-3: Image in a Spoon

A polished soup spoon may be thought of as a concave mirror on one side and a convex mirror on the other side. Based on your observations in this and the last investigations, make the following prediction.

Prediction E3-3: If you look into the concave side of a polished spoon, will your image be upright or inverted? If you look into the convex side, will your image be upright or inverted?

To test your prediction you will need

*   polished soup (or other) spoon
*   light bulb and holder

1. Look into the concave side of the spoon.

Question E3-13: Describe your image. Is it upright or inverted? Enlarged or reduced in size? Do you think it is real or virtual? Explain based on your observations in this lab.

2. Look into the convex side of the spoon.

Question E3-14: Describe your image. Is it upright or inverted? Enlarged or reduced in size? Do you think it is real or virtual? Explain based on your observations in this lab.

If you are not sure of your answers, make the following observations.

3. View the image of a distant light bulb in the concave side of the spoon. Insert a piece of paper close to the spoon, and see if you can locate a real image on the paper.

4. Repeat this procedure with the convex side of the spoon.

Question E3-15: Describe what you observed in each case. Did your observations agree with your answers to Questions E3-13 and E3-14? Explain.

HOMEWORK FOR LAB 4: GEOMETRICAL OPTICS— MIRRORS

1. Define *concave, convex,* and *plane*. Draw sketches of a concave, a convex, and a plane mirror in the space below. (Assume in your sketches that the object is to the left of the mirror.)

2. You use a concave mirror to focus the light from the sun to a small point. Is this point located at the focal point, further from the mirror than the focal point, or closer than the focal point? Explain.

3. Define *focal length*. How can you determine the focal length of a concave mirror?

4. Define *real image* and *virtual image*. What are the similarities and differences between these?

5. A concave mirror has a focal length of 10 cm. Where can you place an object so that you will see a *real* image? Will this image be upright or inverted? In front of the mirror or behind the mirror?

6. What will happen to the image in (5) if you block half of the mirror with a card? If you block half of the object with a card? If you remove the mirror? Explain your answers.

7. For the mirror in (5), where can you place an object so that you will see a *virtual* image? Will this image be upright or inverted? In front of the mirror or behind the mirror?

8. A second concave mirror is curved more so that its radius of curvature is half as large as the radius of curvature of the mirror in (5). What is the focal length of this mirror? Explain.

9. Can a *real* image be formed by a convex mirror? If yes, describe where you should place the object, and where the image will be located. If no, explain why not.

10. Can a real image be formed by a plane mirror? If yes, describe where you should place the object, and where the image will be located. If no, explain why not.

11. What are the similarities and differences between the virtual image formed of an object by a plane mirror and the virtual image of the same object, the same distance away from a convex mirror? Compare (a) the size of the images, (b) the location of the images, and (c) whether the images are upright or inverted.

PRE-LAB PREPARATION SHEET FOR
LAB 5—POLARIZED LIGHT

(Due at the beginning of Lab 5)

Directions:
Read over Lab 5 and then answer the following questions about the procedures.

1. What is waving in an electromagnetic wave?

2. What is polarized light?

3. Why is *wave* optics needed to describe polarized light?

4. What is the rotary motion probe used for in Activity 1-3?

5. How will you find the mathematical relationship between your data from Activity 1-3 for light intensity vs. angle?

6. What do sunglasses have to do with Investigation 2?

LAB 5:
POLARIZED LIGHT

The index of refraction is the tangent of the angle of polarisation . . .
when a ray is polarised by reflection, the reflected ray
forms a right angle with the refracted ray.

—Sir David Brewster, 1815

OBJECTIVES

• To observe the linear polarization of light waves

• To discover different ways of polarizing light

• To find the mathematical law describing the transmission of polarized light through a Polaroid filter—Malus's law

OVERVIEW

As seen in the last several laboratories, the ray model—usually called *ray optics* or *geometrical optics*—is quite adequate to describe the interactions of light with optical elements like lenses. This is because the elements have dimensions much larger than the wavelengths of visible light, and wave interference effects are insignificant.

In this laboratory you will explore one wave aspect of light—*polarization*. In the next lab you will explore other wave behaviors—*interference* and *diffraction*. These phenomena cannot be explained using *geometrical optics*.

Light waves can be described as electromagnetic waves—transverse oscillations of electric and magnetic fields. Polarization refers to the orientation of the electric field. Linearly polarized light always has the electric field vectors of all waves pointing in the same transverse direction, for example, always in the y direction as shown in Figure 5-1 on the following page.

The electric field vectors for unpolarized light point randomly in any transverse direction. (For unpolarized light, the electric field vectors in Figure 5-1, could point in the x direction, the y direction, or any direction in between.)

The wave model of light, called *wave optics* (or *physical optics*), is needed to adequately describe polarization and interference phenomena.

In Investigation 1, you will examine the effect of a Polaroid filter on light. In Investigation 2 you will explore another way to polarize light.

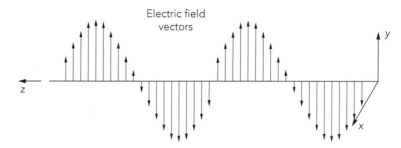

Figure 5-1: Oscillating electric field vectors in an electromagnetic wave.

INVESTIGATION 1: POLARIZING LIGHT WITH A POLARIZING FILTER

Exploring Polarization Using Polaroid Filters

Let's see what effect a Polaroid filter has on light. Since you won't be able to see what is happening to the light waves on a microscopic level, you will have to speculate about just what is going on.

To carry out this activity, you will need the following:

- two Polaroid filter disks with angle markings on them
- light source (e.g., flashlight)

Activity 1-1: Looking at light through a Polaroid filter

1. Set up the light source.
2. Hold one of the Polaroid filter disks in front of your eye, and look directly into the light coming from the light source.
3. Rotate the Polaroid slowly through one full rotation.

Question 1-1: Did anything happen to the light intensity when the Polaroid was placed in front of your eye? Carefully describe what you observed.

Question 1-2: Did anything happen to the light intensity as you rotated the Polaroid? Carefully describe what you observed.

4. Now, mount a Polaroid filter disk on the front of the light source, so that the light from the source passes through the Polaroid. Take the other Polaroid and look through it at the light coming from the source. Rotate the Polaroid in your hand slowly through one complete rotation, as you did in (3).

Question 1-3: Carefully describe what happens now when you rotate the Polaroid in front of your eye.

5. Adjust the Polaroid and light source so that the 0° mark on the Polaroid is straight up. Hold the other Polaroid in front of your eye, beginning with the 0° mark on this Polaroid also pointing straight up. Rotate the Polaroid in front of your eye. Locate all positions of the Polaroid in your hand for which almost all the light gets through, and record these angles below.

    Angles: _____   _____   _____

6. Rotate the Polaroid at right angles to the positions in (5). (Examine all such positions of the Polaroid.)

Question 1-4:   Where is the 0° marker on the Polaroid in your hand pointing when almost all the light gets through? Describe all cases.

Question 1-5:   Describe what happens to the light intensity when the Polaroid is rotated at right angles to the positions in (5).

7. Examine what happens to the light intensity when the Polaroid is rotated to a position in between the positions in (5) and (6).

Question 1-6:   What happens to the light intensity when the Polaroid is in between the positions in (5) and (6)?

A Model for Polarization

A Polaroid filter is a piece of plastic with special molecules oriented on its surface. The direction of the oriented molecules defines the *axis* of the Polaroid. The axis is along a line drawn from 0° to 180° on your Polaroid disks. Suppose that according to our model

  a. Light waves with their electric field vectors pointing parallel to the axis are almost completely transmitted through the Polaroid (Figure 5-2).

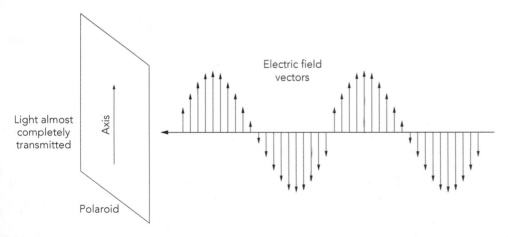

Figure 5-2: Orientation of Polaroid axis for which vertically polarized light is almost completely transmitted.

b. Light waves with their electric field vectors pointing perpendicular to the axis are not transmitted by the Polaroid (Figure 5-3).

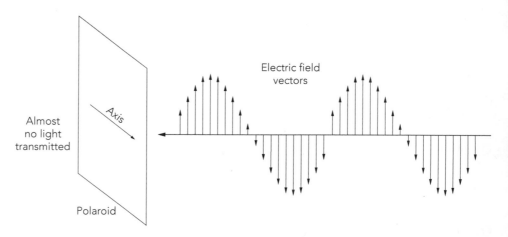

Figure 5-3: Orientation of the Polaroid axis for which almost no vertically polarized light is transmitted.

Now try to explain your observations using this model for polarization.

Activity 1-2: Explaining Your Observations of Polarization

Question 1-7: Consider the light source without the Polaroid mounted in front of it. Based on this model and your observations in Activity 1-1, what can you say about the orientation of the electric field vectors of the waves coming from the light source?

Question 1-8: Now the Polaroid is mounted in front of the light source. (This is called the *polarizer*.) What can you now say about the orientation of the electric field vectors of the light coming from the light source after passing through the polarizer?

Question 1-9: Explain your observations when looking at this light through the second Polaroid (called the *analyzer*) and rotating it. What is going on when the axes of the two Polaroids are parallel? What is going on when the axes of the two Polaroids are perpendicular?

Question 1-10: Considering that the electric field is a vector, explain the variation of intensity when the Polaroids are between parallel and perpendicular. (*Hint:* Consider Figures 5-2 and 5-3. What would happen if the Polaroid were rotated so that its axis made an angle between 0° and 90° with the direction of the electric field?)

A Quantitative Look

You can now look at the behavior of linearly polarized light passing through a Polaroid filter more quantitatively. Your goal is to find a mathematical relationship that predicts how much light gets through the Polaroid filter (the *analyzer*) as a function of the angle between the axis of polarization of the light (the direction along which the electric field points) and the axis of the analyzer.

Prediction 1-1: First sketch your prediction of the intensity transmitted as a function of angle on the axes that follow. Base your prediction on your observations and explanations in Activities 1-1 and 1-2.

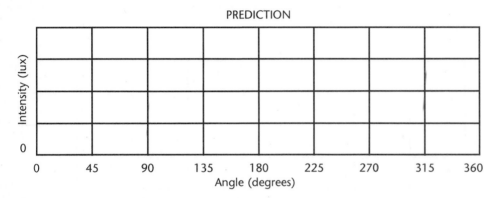

To test your prediction you will need

- computer-based laboratory system
- *RealTime Physics Light and Optics* experiment configuration files
- light probe
- rotational motion probe
- stand
- small Polaroid disk with angle markings
- large Polaroid disk
- light source (e.g., flashlight)
- small lens and holder (optional)

Activity 1-3: Dependence of Transmitted Intensity on Angle

1. Set up the equipment as shown in Figure 5-4.

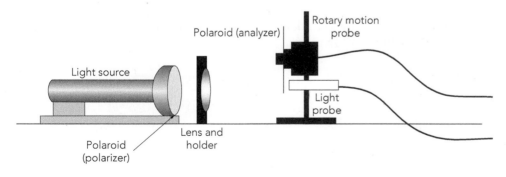

Figure 5-4: Experimental setup for analyzing polarized light (view from the side).

The light probe should be positioned exactly opposite the light source and right behind the large Polaroid disk. The rotational motion probe should be mounted on the stand, so that it is suspended above the table and free to rotate.

The polarizer should have the 0° mark pointing straight up. The lens (if needed) should be placed so that light from the source has not spread out very much when it reaches the light probe.

2. Plug the light probe and rotary motion probe into the interface.

3. Open the experiment file called **Intensity vs. Angle (L05A1–3).** The axes that follow will be displayed. You will want to graph intensity measured by the light probe vs. angular position of the rotational motion probe in degrees for one full rotation (360°).

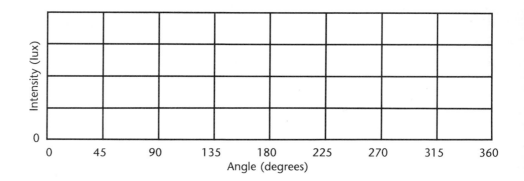

4. Be sure that the light probe is not saturating. Rotate the analyzer through a complete turn, and be sure that the Intensity reading does not reach a maximum value where it remains as you rotate further. If the probe is saturating, move the light source and lens further away, or change the range setting on the light probe until the probe reading varies continuously as the analyzer is rotated through a complete turn.

5. **Adjust** the Intensity scale so that its maximum value includes the maximum intensity measured by the light probe.

6. You are now ready to collect data. Be sure that to begin with the analyzer is at the position that lets maximum intensity through to the light probe. Then **begin graphing.**

When the intensity numbers appear on the screen, begin rotating the analyzer at a slow, uniform rate until you complete one revolution. You can then **stop graphing.**

7. **Print** the graph and affix it over the axes above.

Question 1-11: Does the variation in intensity as the analyzer is rotated agree with your Prediction 1-1? Explain.

Prediction 1-2: Based on your graph, what would you say is the mathematical dependence of the intensity transmitted by the analyzer as a function of the angle between the direction of polarization of the incident light and the direction of the axis of the analyzer? In your equation, use I as the transmitted intensity, I_0 as the maximum transmitted intensity, and θ as the angle. Explain how you arrived at your prediction.

You can test your prediction by creating a mathematical model with the software.

Activity 1-4: Mathematical Model for Transmitted Intensity as a Function of Angle

You may display a graph for any mathematical function of angle by creating a **calculated column** in the software.

1. Create a **calculated column** for Intensity as a function of the measured angle. **Display** both the measured intensity from Activity 1-3 and your mathematical model on the same graph axes.

2. Try functions and change the parameters until you get a mathematical model that clearly matches your measured Intensity vs. Angle data.

   Hint 1: Your measured angles are in degrees, but, in your software, the trigonometric functions may assume that the angle is measured in radians. Be sure to add whatever factors are necessary in your function to correct for this.

   Hint 2: You may need to insert an additive phase factor to get the two graphs to be exactly on top of each other.

   Hint 3: Try trigonometric functions or trigonometric functions raised to a power.

3. When you have gotten the best possible match, **print** your graphs and affix them in the space below. Be sure that the equation of your model is also displayed, or write it in on your graph.

Question 1-12: What do you conclude is the mathematical relationship between intensity of polarized light transmitted through a Polaroid and angle between the axis of polarization of the light and the axis of the Polaroid? Explain how you reached your conclusion.

Question 1-13: Did your mathematical relationship for Intensity vs. Angle agree with your Prediction 1-2? Explain.

Question 1-14: Did your mathematical model for Intensity vs. Angle match your experimental data well? Explain.

Question 1-15: Based on the fact that the intensity of a light wave is proportional to the square of the electric field magnitude, and your answers to the previous questions, explain why the transmitted intensity should have this mathematical dependence on the angle.

Comment: The observed dependence of the transmitted intensity on the angle between the axis of polarization and the axis of the Polaroid, $I = I_0 \cos^2\theta$, is known as Malus's law.

INVESTIGATION 2: OTHER WAYS OF POLARIZING LIGHT

Polaroid Sunglasses

When you're lying on the beach on a hot sunny day, have you wondered how Polaroid sunglasses cut down the glare of sunlight reflected from the surface of the water? Or how they block light reflected from the windshield of the car driving ahead of you? In the previous activities you found that unpolarized light becomes polarized by passing through a Polaroid filter. In this investigation, you will explore another way for light to become polarized.

You will need the following equipment:

- computer-based laboratory system
- *RealTime Physics Light and Optics* experiment configuration files
- light probe
- rotational motion probe
- stand
- large Polaroid disk
- glass plate and holder
- light source (e.g., flashlight)
- small lens and holder

Activity 2-1: Analyzing Reflected Light

In this activity you will analyze light that is reflected off a glass surface to see if it is polarized.

1. Set up the equipment as shown in Figure 5-5. Be sure to remove the polarizer from the light source, and rearrange the equipment as shown. The light is now reflected by a glass plate toward the light probe.

    Adjust the position of the lens so that the beam doesn't spread out very much in reaching the light probe.

2. Open the experiment file called **Reflected Intensity (L05A2-1).** The axes that follow will be displayed. You will want to graph intensity measured by the light probe vs. angular position of the rotational motion probe in degrees for one full rotation (360°).

3. Be sure that the light probe is not saturating. Rotate the analyzer through a complete turn, and be sure that the Intensity reading does not reach a maximum value where it remains as you rotate further. If the probe is saturating, move the light source and lens, or change the range setting on the light probe until the probe reading varies continuously as the analyzer is rotated through a complete turn.

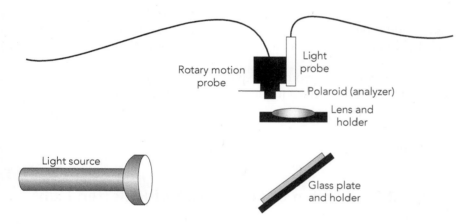

Figure 5-5: Experimental setup for analyzing reflected light (viewed from above).

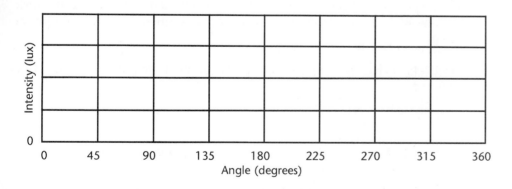

4. **Adjust** the Intensity scale so that its maximum value includes the maximum intensity measured by the light probe.

5. You are now ready to collect data. Be sure that to begin with the analyzer is at the position that lets maximum intensity through to the light probe. Then

begin graphing, and graph intensity vs. angular position as you rotate the analyzer through one complete rotation. **Stop graphing.**

6. **Print** your graph and affix it over the axes above.

Question 2-1: Describe any variation of intensity as the analyzer was rotated. Did this variation resemble in any way your observations in Investigation 1?

Question 2-2: When unpolarized light is incident on a glass plate, and reflected from it, what seems to happen to the light as far as its polarization? Explain.

Question 2-3: Can you explain on the basis of your graph why you might want to make sunglasses out of Polaroid material? Under what circumstances would these sunglasses be especially effective?

As you have observed, when light is reflected off the surface of a transparent material like glass, the reflected light is polarized. To be more quantitative, two questions should be answered: (1) in what direction is the axis of polarization of the reflected light, and (2) does the degree of polarization depend on the angle of incidence?

If you have more time, you can explore these two questions in the following extensions. You will need the equipment above and the following:

- protractor
- small Polaroid disk with angle markings on it

Extension 2-2: Direction of Polarization of Reflected Light

1. Use the same setup as in Activity 2-1. You can remove the light probe and rotary motion probe and use your eyes to make qualitative observations.

2. Look through the small Polaroid disk at the reflected light. Rotate the disk and determine the angle(s) for which the transmitted light has minimum intensity.

Question E2-4: What is the direction of polarization of the reflected light? How do you know?

Question E2-5: How is the direction of polarization of the reflected light related to the plane of incidence? (Recall from Lab 2 that the *plane of incidence* contains the incident ray and the normal to the glass plate's surface.) Is the direction of polarization in the plane of incidence, perpendicular to it, or something else? Explain.

Extension 2-3: Brewster's Angle

Does the angle of incidence make any difference in the degree of polarization of the light? You can examine this question with the following procedure.

1. Use the same setup as in Extension 2-2. You can change the angle of incidence by rotating the glass plate and its holder.

2. View the reflected light with the small Polaroid rotated to transmit the minimum intensity. Rotate the glass plate horizontally, and observe what happens to the transmitted intensity. Be sure to keep the small Polaroid centered on the light reflected from the glass plate.

Question E2-6: Did you observe any change in the transmitted intensity as the glass plate was rotated? Could this mean that the light was not as completely polarized by reflection at some angles? Explain.

3. Find the position of the glass plate for which you observe the smallest transmitted intensity with the Polaroid at the same angle as in (2). (Find the position of the glass plate for which there is the greatest degree of polarization of the light that is reflected by the plate.)

4. Use the protractor to measure the angle of incidence for this position of the glass plate.

    Angle of incidence for maximum polarization: _____

Comment: According to theory and experiment, the reflected light is *completely polarized* only when the angle of incidence is given by

$$\tan \theta_B = n$$

where n is the index of refraction of the glass and θ_B is called Brewster's angle, which is characteristic of the reflecting material.

Question E2-7: Use the index of refraction to determine Brewster's angle for the glass you used to polarize the light. Show your calculation.

Question E2-8: How does the value you calculated in Question E2-7 compare with the angle of incidence that you measured? How well could you measure the angle? Do the two angles agree within this uncertainty?

Name_____ Date_____ Partners_____

HOMEWORK FOR LAB 5: POLARIZED LIGHT

1. What is polarized light? How would you determine if light from some source is polarized?

2. Describe two ways of polarizing light that you observed in this laboratory.

3. You view light from a flashlight through a Polaroid. Describe any intensity variation as the Polaroid is rotated. Explain your answer.

   Describe any intensity variation as the flashlight is rotated.

4. Unpolarized light travels through two pieces of Polaroid as shown below.

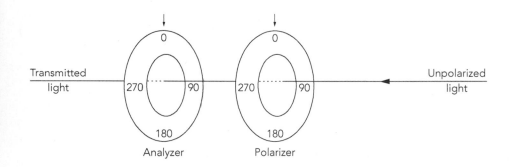

   When the two Polaroids are rotated to 0° as shown, the transmitted intensity is I_0. What percentage of I_0 will be transmitted when the analyzer is rotated to 90°? Explain.

   What percentage of I_0 will the transmitted intensity be if the analyzer is rotated to 30°? Show your calculation.

5. Light is reflected off the rear windshield of a car.

When the Polaroid is rotated so that the 0° or 360° mark is at the top, the transmitted intensity is minimum, while at 90° and 270° it is maximum. Explain why this variation in intensity occurs.

6. What is the direction of the axis of polarization of the light reflected from the windshield in (5)—up and down or side-to-side?

If you were designing Polaroid sunglasses, in which of the following direc-

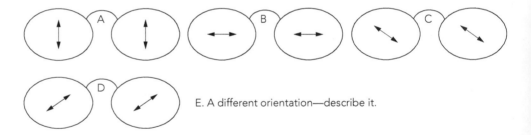

E. A different orientation—describe it.

tions should the axis of the Polaroid point? Explain your answer.

7. Lucite has index of refraction 1.49. What is Brewster's angle for Lucite? Show your calculation.

If unpolarized light is incident on the surface of a piece of Lucite at this angle, describe the degree of polarization of the reflected light.

PRE-LAB PREPARATION SHEET FOR
LAB 6—WAVES OF LIGHT

(Due at the beginning of Lab 6)

Directions:
Read over Lab 6 and then answer the following questions about the procedures.

1. What type of light source will be used in this lab?

2. In words, what is your Prediction 1-1?

3. What is different about the observations you will make in Activity 2-1 and those you made in Activity 1-2?

4. What will the rotary motion probe be used for in this lab?

5. For what size objects must we use a wave model to correctly describe the interactions of light with the objects?

LAB 6:
WAVES OF LIGHT

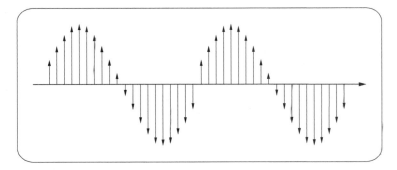

Whenever two portions of the same light arrive at the eye by different routes, either exactly or very nearly in the same direction, the light becomes most intense where the difference of the routes is a multiple of a certain length, and least intense in the intermediate state for the interfering portions; and this length is different for light of different colours.

—Thomas Young, 1802

OBJECTIVES

- To observe how light waves behave when they pass through narrow slits

- To explore light interference in general

- To understand the light pattern that results when light waves from different parts of the same narrow slit interfere with each other

- To understand the light pattern that results when light waves from two parallel narrow slits interfere with each other

OVERVIEW

In Lab 5 you investigated polarization, a phenomenon that can only be described by a *wave model* for light. In Labs 1–4 you examined interactions of light with optical elements like lenses and mirrors that have dimensions much larger than the wavelength of the light. However, in this lab you will see what happens when light interacts with much smaller optical elements, like narrow slits, with sizes no larger than a small number of wavelengths. Again, as with polarization, the ray model—*ray optics* or *geometrical optics*—is completely inadequate in describing these interactions of light.

We will study the resulting wave behaviors called *interference* and *diffraction*. As described in the quotation above by Thomas Young, wave interference effects occur when a light wave is separated into two or more parts (e.g., by passing through two small parallel slits), and the parts are later brought together. Interesting and often beautiful interference patterns are formed through the superposition (adding together) of these waves.

The wave model of light, called *wave optics* (or *physical optics*) is needed to adequately describe *interference* and *diffraction*. In general, *interference* describes the results when a wave front is divided into parts by passing through two or more distinct holes on slits, while *diffraction* describes what happens when extended light waves are blocked or restricted by a single object or slit. In this case, interference takes place between waves originating from different points on the same wave front.

INVESTIGATION 1: SINGLE-SLIT DIFFRACTION

You will first observe what happens when light passes through a single slit. To begin with, try a slit that is very large compared to the wavelength of light. A common, red, helium–neon laser emits light with wavelength 632.8 nm (632.8 × 10^{-9} m = 0.0000006328 m).

Question 1-1: How many wavelengths of a helium–neon laser are equivalent to 1.0 mm? Show your calculation.

Prediction 1-1: Suppose that light from the helium–neon laser is aimed through a 1-mm-wide slit as shown in Figure 6-1. Based on the *ray* or *geometrical optics* model, what would you see on a screen 1.5 meters away? Sketch your prediction and describe below in words how the light on the screen would be different with and without the slit.

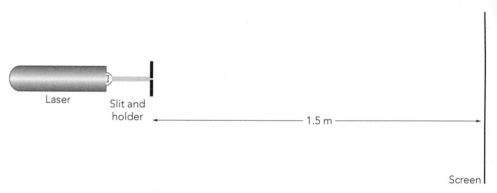

Figure 6-1: Experimental setup for observing laser light through a slit. (View is from above.)

To test your prediction, you will need the following:

- 1-mm-wide slit and holder
- helium–neon laser
- screen
- blackboard eraser and chalk dust

Activity 1-1: Light Through a Slit

1. Set up the laser, slit, and screen with the slit vertical, as in Figure 6-1.

2. Turn on the laser, and observe the light on the screen with and without the slit. You will be able to see the path of the light better if you spread chalk dust in the region between the slit and the screen.

Question 1-2: Was your prediction correct? Describe what you observed. Does ray optics give a correct prediction for this observation?

Now, suppose the slit were much narrower—nearer in size to the wavelength of the laser light. Let's first examine qualitatively what happens when the light passes through a very narrow slit.

In addition to the equipment above, you will need

*   slit plate and holder
*   magnifying lens

Activity 1-2: Single-Slit Diffraction

1. Examine two different single slits on the slit plate through the magnifying lens. Call these slits 1 and 2.

Question 1-3: Describe the slits. How are they different from each other? How are they different from the 1-mm slit you used in Activity 1-1?

2. According to the data for the slit plate, how wide are slits 1 and 2?

    Slit 1 width:_____   Slit 2 width:_____

Question 1-4: How many wavelengths of a helium–neon laser are equivalent to the widths of slits 1 and 2? Show your calculations.

    Slit 1 width:_____   Slit 2 width:_____

3. Replace the 1-mm slit by the slit plate, and align the plate so that the laser is shining through single slit 1, which is vertical.

4. Again spread chalk dust between the slit and the screen.

Question 1-5: Carefully describe what you observe on the screen, and compare it to what you saw with the 1-mm slit in Activity 1-1.

5. Replace slit 1 with slit 2. Carefully compare the patterns for slits 1 and 2. If necessary, switch back and forth several times.

Question 1-6: Carefully describe what is similar and what is different about the patterns with slits 1 and 2.

Now you can examine the light intensity variation along the screen using the light probe and rotary motion probe. First a prediction.

Prediction 1-2: Look at Figure 6-2. As you have seen in previous labs, a light probe measures light intensity. Suppose that the light probe is moved across the position of the screen. Based on your qualitative observations in this activity, sketch on the axes below the intensity of light vs. distance across the screen. Label your graph slit 1. Also explain your prediction in words below the graph.

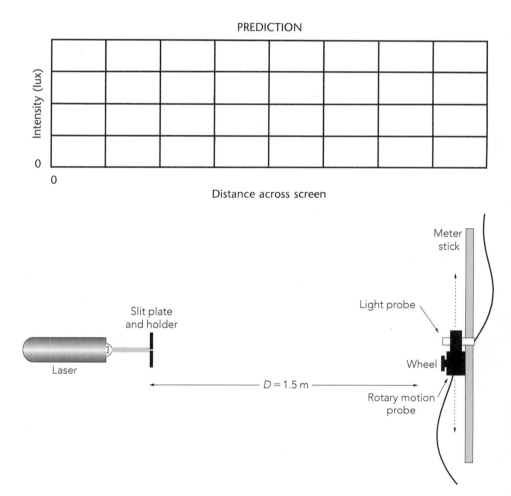

Figure 6-2: Experimental setup for observing laser light through a narrow slit quantitatively. (View is from above.)

To test your prediction, you will need the following:

- helium–neon laser
- slit plate and holder

- height adjusters for the laser and slit holder

- computer-based laboratory system

- *RealTime Physics Light and Optics* experiment configuration files

- light probe

- rotary motion probe and meter stick or track

- table clamps

Activity 1-3: Single-Slit Diffraction—A Quantitative Look

1. Set up the laser and slit and holder on the height adjusters, so that the laser beam and slit can be raised and lowered independently.

2. Set up the light probe, rotary motion probe, and meter stick in place of the screen as shown in Figure 6-2. The meter stick should be fastened in place perpendicular to the laser beam using the table clamps. The box of the rotary motion probe should be on the table, against the meter stick. The probe should be free to move along its wheel across the light pattern from the laser and slit.

3. Set the light probe to its most sensitive range.

4. Open the experiment file **Intensity vs. Position (L06A1-3).** This should open the axes for intensity vs. position that follow.

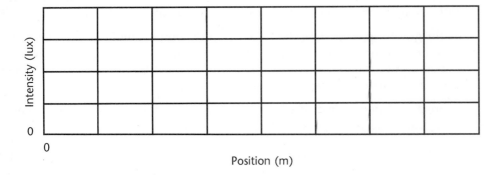

5. Move the rotary motion probe so that the light probe is at the middle of the light pattern. Adjust the alignment of the laser beam and slit plate so that the beam is centered on the light probe. You can fine tune these adjustments by maximizing the light probe reading.

6. **Set the Intensity axis** so that the maximum value is larger than the maximum value you measured at the center of the light pattern.

7. Measure the distance, D, from the slit plate to the light probe photodiode. It should be close to 1.5 m.

   D:_____

8. Slide the rotary motion probe back and forth, all the way across the track, to see that the light is centered on the light probe all the way across the light pattern. If the light pattern is not level, then rotate the slit plate slightly until the pattern is level and hits the center of the probe all the way across.

Comment: The rotary motion probe can be calibrated to read linear position instead of angular position by recording the diameter of the wheel in the software. The diameter may be most accurately measured by sliding the rotary motion probe through a measured linear distance that corresponds to one rotation of the wheel. Then the linear distance S for one rotation is equal to the circumference, $S = \pi d$ and $d = S/\pi$.

9. Slide the rotary motion probe all the way to one end of the track. **Begin graphing,** and slide it—applying pressure so that the wheel doesn't slip— until it has rotated exactly one rotation. (If no rotation is displayed on the graph, you may need to **change the positive rotation direction** in the software.) Measure S, and calculate d. Show your calculation.

   S:_____   d:_____

10. Set up the software to measure linear position in cm by **entering** your value for the diameter, d, of the wheel in cm.

11. Move the rotary motion probe so that the extreme left edge of the laser light pattern is shining on the light probe. Mark this position on the meter stick. Now move the probe to the extreme right end of the light pattern. Determine the width across the entire light pattern, and record it below.

   Width of light pattern:_____cm

12. **Set the Position axis** on your graph in the software so that it is at least as large as the width of the light pattern.

13. You are now ready to collect data. Begin with the rotary motion probe aligned with the mark you made at the extreme left side of the light pattern.

14. **Begin graphing,** and slide the rotary motion probe slowly and steadily to the other end of the light pattern. **Stop graphing.** Repeat if necessary to get a good graph. When you have a good graph, adjust the software so that the graph is **persistently displayed** on the screen.

15. Change the **Intensity** axis if necessary to better display your data. Then **print** your graph and affix it over the axes above.

Question 1-7: Describe the intensity vs. position graph in words. Is the observed graph what you expected? Explain how it corresponds to what you observed on the screen in Activity 1-2.

Question 1-8: Describe in words what happens to the light when it passes through a narrow slit.

Comment: As you have seen in previous labs, light from a laser is special. It has a very narrow wavelength range, unlike white light (we say that it is very nearly *monochromatic*), and it doesn't spread out very much in space. One other property is that, to a very high degree, all of the waves emitted by a laser start out in step (in *phase*) with each other. We say that the light is *coherent*. When light with these properties is restricted by a narrow slit, the result is the pattern that you have observed both qualitatively and quantitatively. This is called a *single-slit diffraction pattern*.

Extension 1-4: Single-Slit Diffraction with a Different Slit

Prediction E1-3: Suppose that you replaced slit 1 by slit 2 in Figure 6-2. Based on your qualitative observations in Activity 1-2, sketch on the Prediction axes above the new graph of intensity of light vs. distance across the screen. Label your graph Slit 2. Also explain your prediction in words below, noting any significant differences from the graph you observed in Activity 1-3.

Test your prediction.

1. Replace slit 1 with slit 2, and follow the steps in Activity 1-3 to carefully re-align the laser, slit, rotary motion probe, and light probe.

2. Keep the graph from Activity 1-3 **persistently displayed** on the screen.

3. Position the rotary motion probe at the same starting point that you marked on the meter stick in Activity 1-3.

4. **Begin graphing,** and slide the rotary motion probe slowly and steadily to the other end of the light pattern. **Stop graphing.** Repeat if necessary to get a good graph.

5. Change the **Intensity** axis if necessary to better display your data. Then **print** your graph and affix it over the axes above.

Question E1-9: Did your graph agree with your prediction? Explain how any differences from the graph for Activity 1-3 correspond to qualitative differences that you observed on the screen in Activity 1-2.

Question E1-10: Is the diffraction pattern wider or narrower for the narrower slit? Explain based on your observations.

INVESTIGATION 2: TWO-SLIT INTERFERENCE

As you have seen in the previous investigation, when *coherent* light waves pass through a slit that is not too wide compared to the wavelength of the light, the resulting *single-slit diffraction* pattern is very different from what a ray model (geometrical optics) would predict. Now let's see what happens when light waves go through two narrow parallel slits separated by a distance that is small compared to the wavelength.

You will need the same equipment as in Activity 1-3:

- helium–neon laser
- slit plate and holder
- height adjusters for the laser and slit holder

- computer-based laboratory system
- *RealTime Physics Light and Optics* experiment configuration files
- light probe
- rotary motion probe and meter stick or track
- table clamps
- screen
- magnifying lens
- blackboard eraser and chalk dust

Activity 2-1: Light Through Two Slits

1. Examine two different sets of double slits on the slit plate through the magnifying lens. The slits should have the same width as the slits in Investigation 1, but different separations. Call these slits 1 and slits 2.

Question 2-1: Describe the slits. How are they different from each other? How are they the same?

2. According to the data for the slit plate, how wide are slits 1 and slits 2? What are their separations?

Slits 1 width:_____   Slits 2 width:_____

Slits 1 separation:_____   Slits 2 separation_____

Question 2-2: How many wavelengths of a helium–neon laser are equivalent to the separation of slits 1 and slits 2? Show your calculations.

Slits 1 separation:_____   Slits 2 separation:_____

3. Use the same setup as in Figure 6-2, but position slits 1 in front of the laser. Put the screen in front of the rotary motion probe so that you can observe the light pattern qualitatively. Be careful not to move the ruler and rotary motion probe setup.

4. Again spread chalk dust in the region between the slit and the screen.

Question 2-3: Carefully describe what you observe on the screen, and compare it to what you saw in Investigation 1 with a single slit.

Question 2-4: If ray (geometrical optics) accurately described this situation, what would the pattern on the screen look like? Explain.

5. Replace slits 1 with slits 2. Carefully compare the patterns for slits 1 and slits 2. If necessary, switch back and forth several times.

Question 2-5: Carefully describe what is similar and what is different about the patterns with slits 1 and slits 2.

Now you can examine the light intensity variation along the screen as in Investigation 1 using the light probe and rotary motion probe. First a prediction.

Prediction 2-1: Look at Figure 6-2. Suppose that the light probe is moved across the position of the screen. Based on your qualitative observations in this activity, sketch on the axes below the intensity of light vs. distance across the screen for the double slits 1. Label your graph Slits 1. Also explain your prediction in words below the graph.

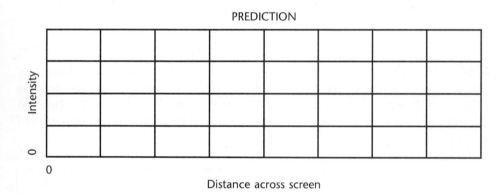

Activity 2-2: Two-Slit Interference

1. Use the same experiment file as in Activity 1-3, and keep the single-slit pattern from Activity 1-3 **persistently displayed** on the screen. (If the light pattern from Extension 1-4 is also displayed, **hide** or **delete** it.)

2. Follow the steps in Activity 1-3 to carefully realign the laser, slit, rotary motion probe, and light probe with slits 1 in front of the laser.

3. Position the rotary motion probe at the same starting point that you marked on the meter stick in Activity 1-3.

4. **Begin graphing,** and slide the rotary motion probe slowly and steadily to the other end of the light pattern. **Stop graphing.** Repeat if necessary to get a good graph.

5. **Adjust** the Intensity axis if necessary to better display your data. Then **print** your graph and affix it in the space below.

6. Use the **analysis feature** of the software to find the positions of the *centers* of all of the peaks in the two-slit pattern. Record these data on your graph, for use in the following extension.

Question 2-6: Is the intensity vs. position graph what you expected for the two slits? Explain how it corresponds to what you observed on the screen in Activity 2-1.

Question 2-7: How does the light pattern for the two slits compare to the single-slit diffraction pattern from Activity 1-3? Carefully describe any similarities and any differences between the two patterns displayed on the computer screen.

Question 2-8: Does there appear to be a relationship between the maximum intensity of the double-slit bright spots (this activity) and the intensity pattern of a single slit of the same width (Activity 1-3)? Do the heights of the peaks observed here seem to follow the same pattern as the intensity of the single-slit pattern?

> **Comment:** When a light wave goes through two slits, it is separated into two parts. The light pattern on the screen is called a *two-slit interference pattern*. It is determined by the way in which these two waves add together by *superposition*. At points on the screen where they are in phase, they add constructively and the intensity is *maximum*. At points where they are exactly out of phase, they add destructively and the intensity is minimum. As you have observed, there are a number of places on the screen where each of these conditions occur.

If you have more time, carry out the following extensions.

Extension 2-3: A More Quantitative Look

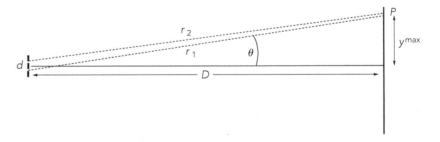

Figure 6-3: Geometry of a two-slit interference pattern.

As you can see in Figure 6-3, the waves from the top slit travel a shorter distance, r_2, than the distance, r_1, traveled by waves from the bottom slit in reaching the same point P on the screen. If the difference in path is a whole number of wavelengths, then the two waves will be in step when they reach the screen and will add constructively, resulting in a bright spot. Using trigonometry, it is possible to show that the condition is

$$r_1 - r_2 = d \sin \theta = m\lambda$$

where m is the number of the bright spot on the screen counted in both directions from the center (brightest) one for which $m = 0$. m is called the *order of interference.*

In Figure 6-3, you can also see that

$$y^{max} = D \tan \theta$$

1. Use the data you recorded on your graph for the positions of the center bright spot and the third bright spot ($m = 3$) on either side of it to find an average value for y^{max}. Show your calculation.

   Positions:

   Center bright spot:_____ cm $m = 3$ bright spot:_____ cm

   Average y^{max}:_____ cm

2. Use this value, the wavelength of the laser light (632.8 nm), and D, the distance from the slits to the light probe, to find the slit spacing, d. Show your calculations below. Watch your units!

   d:_____ cm   (slits 1)

Question E2-9:   Compare your value for d to the value given by the manufacturer of the slits.

Extension 2-4: Interference Pattern for a Different Pair of Slits

Prediction E2-2:   Suppose that you replaced slits 1 by slits 2 in Figure 6-2. Based on your qualitative observations in Activity 2-1, sketch on the Prediction axes above the new graph of intensity of light vs. distance across the screen. Label your graph slits 2. Also explain your prediction in words below, noting any significant differences from the graph you observed in Activity 2-2.

1. Use the same experiment file as in Activity 2-2, and keep the two-slit pattern from Activity 2-2 **persistently displayed** on the screen.

2. Follow the steps in Activity 1-3 to carefully realign the laser, slits, rotary motion probe, and light probe with slits 2 in front of the laser.

3. Position the rotary motion probe at the same starting point that you used in Activities 1-3 and 2-2.

4. **Begin graphing,** and slide the rotary motion probe slowly and steadily to the other end of the light pattern. **Stop graphing.** Repeat if necessary to get a good graph.

5. **Adjust** the Intensity axis if necessary to better display your data. Then **print** your graph and affix it in the space below.

6. Use the **analysis feature** of the software to find the positions of the *centers* of all of the peaks in the two-slit pattern. Record these.

Question E2-10: How does your graph compare to your prediction? Describe the similarities and differences between this graph and the one for Activity 2-2.

Question E2-11: Based on your observations here and in Activity 2-2, how does changing the spacing between the slits affect the observed interference pattern?

Extension 2-5: A More Quantitative Look

Use the method of Extension 2-3 to find the slit spacing for slits 2.

1. Use the data you recorded on your graph for the positions of the center bright spot and the third bright spot ($m = 3$) on either side of it to find an average value for y^{max}. Show your calculation.

   Positions:

   Center bright spot:_____ cm $m = 3$ bright spot:_____ cm

   Average y^{max}:_____ cm

3. Use this value, the wavelength of the laser light (632.8 nm), and D, the distance from the slits to the light probe, to find the slit spacing, d. Show your calculations below. Watch your units!

   d:_____ cm   (slits 2)

Question E2-12: Compare your value for d to the value given by the manufacturer of the slits.

HOMEWORK FOR LAB 6: WAVES OF LIGHT

1. What are *interference* and *diffraction*? Under what circumstances do they occur?

2. Light from a helium–neon laser is shined on a slit of width 0.02 mm. Describe *qualitatively* the pattern of light seen on a screen 1.0 meter away.

3. How would the pattern in (2) be different if the slit were 0.06 mm wide instead of 0.02 mm?

4. How would the pattern in (2) be different if instead of one slit, there were two slits, each 0.02 mm wide, separated by 0.2 mm?

5. How would the pattern in (4) be different if there were two slits, each 0.02 mm wide, separated by 0.4 mm?

6. How does the width of the slits affect the intensity pattern of a two-slit interference pattern?

Questions 7–9 refer to the graph of light intensity vs. position of the probe shown below. The graph was made with helium–neon laser light (632.8 nm) shined through two very narrow slits separated by a small distance. The slits were 2.0 meters away from the probe.

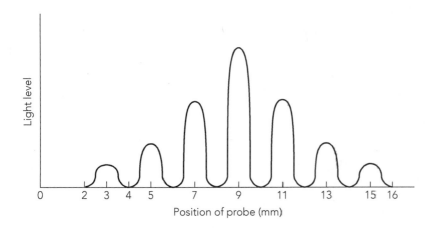

7.   What is the separation of the slits? Show your calculations.

8.   What is the width of the slits? Show your calculations.

9.   Sketch on the same axes the approximate light intensity pattern you would get if the slits were twice as far apart. (Assume that the slit width is unchanged, and try to get the intensity maxima at the right location.)

Listed below are the settings in the *Experiment Configuration Files* used in these labs. These files are available from Vernier Software and Technology for *Logger Pro* software (Windows and Macintosh) and from PASCO for *Data Studio* (Windows and Macintosh). They are listed here so that the user can set up files for any compatible hardware and software package.

Experiment file	Description	Data collection	Data handling	Display
Point Source (L0A1-1a)	Displays Intensity in lux on a meter.	Light probe	NA	Meter with Intensity in lux.
Intensity vs. Distance (L01A1-1b)	Displays data table and axes for Intensity vs. Distance. Data can be entered into the table, and an active graph results.	NA Data are entered manually in the data table.	Fit routine used to find relationship between Intensity and Distance.	One set of graph axes. Intensity: 0–600 lux (adjustable). Distance: 0–40 cm.
Reflected Light (L01A2-1)	Displays Intensity in lux on a meter.	Light probe.	NA	Meter with Intensity in lux.
Light from Objects (L01A2-2)	Displays and graphs Intensity vs. Time.	50 points/sec. Intensity from light probe; graphical display of input.	NA	One set of graph axes with line. Intensity: 0–600 lux (adjustable). Time: 0–20 sec.
Snell's Law (L02A1-2)	Graphs $\sin\theta_1$ vs. $\sin\theta_2$ for values entered manually.	NA Data entered manually in data table.	Fit routine used to find relationship between $\sin\theta_1$ and $\sin\theta_2$.	One set of graph axes with line. $\sin\theta_1$: 0–1 $\sin\theta_2$: 0–1
Intensity vs. Angle (L05A1-3)	Displays and graphs Intensity2 vs. Angle1.	50 points/sec. Rotary motion probe 1 and Light probe 2. Digital and graphical display of inputs. Data collection for 50 sec.	NA	One set of graph axes with line. Angle: 0–400 deg. Intensity: 0–600 lux (adjustable).
Reflected Intensity (L05A2-1)	Displays and graphs Intensity2 vs. Angle1.	50 points/sec. Rotary motion probe 1 and Light probe 2. Digital and graphical display of inputs. Data collection for 50 sec.	NA	One set of graph axes with line. Angle: 0–400 deg. Intensity: 0–600 lux (adjustable).
Intensity vs. Position (L06A1-3)	Displays and graphs Intensity2 vs. Linear Position1.	50 points/sec. Rotary motion probe 1 and Light probe 2. Digital and graphical display of inputs. Diameter of wheel is entered so display of rotary motion probe is in linear distance. Data collection for 50 sec.	NA	One set of graph axes with line. Linear distance: 0–50 cm. Intensity: 0–100 lux (adjustable).